Mathematical Delights

© 2004 by
The Mathematical Association of America (Incorporated)
Library of Congress Catalog Card Number 2004100962

ISBN 088385-334-5

Printed in the United States of America

Current printing (last digit):
10 9 8 7 6 5 4 3 2 1

The Dolciani Mathematical Expositions

NUMBER TWENTY-EIGHT

Mathematical Delights

Ross Honsberger
University of Waterloo

Published and distributed by
The Mathematical Association of America

DOLCIANI MATHEMATICAL EXPOSITIONS

The DOLCIANI MATHEMATICAL EXPOSITIONS series of the Mathematical Association of America was established through a generous gift to the Association from Mary P. Dolciani, Professor of Mathematics at Hunter College of the City University of New York. In making the gift, Professor Dolciani, herself an exceptionally talented and successful expositor of mathematics, had the purpose of furthering the ideal of excellence in mathematical exposition.

The Association, for its part, was delighted to accept the gracious gesture initiating the revolving fund for this series from one who has served the Association with distinction, both as a member of the Committee on Publications and as a member of the Board of Governors. It was with genuine pleasure that the Board chose to name the series in her honor.

The books in the series are selected for their lucid expository style and stimulating mathematical content. Typically, they contain an ample supply of exercises, many with accompanying solutions. They are intended to be sufficiently elementary for the undergraduate and even the mathematically inclined high-school student to understand and enjoy, but also to be interesting and sometimes challenging to the more advanced mathematician.

———

1. *Mathematical Gems,* Ross Honsberger
2. *Mathematical Gems II,* Ross Honsberger
3. *Mathematical Morsels,* Ross Honsberger
4. *Mathematical Plums,* Ross Honsberger (ed.)
5. *Great Moments in Mathematics (Before 1650),* Howard Eves
6. *Maxima and Minima without Calculus,* Ivan Niven
7. *Great Moments in Mathematics (After 1650),* Howard Eves
8. *Map Coloring, Polyhedra, and the Four-Color Problem,* David Barnette
9. *Mathematical Gems III,* Ross Honsberger
10. *More Mathematical Morsels,* Ross Honsberger
11. *Old and New Unsolved Problems in Plane Geometry and Number Theory,* Victor Klee and Stan Wagon
12. *Problems for Mathematicians, Young and Old,* Paul R. Halmos
13. *Excursions in Calculus: An Interplay of the Continuous and the Discrete,* Robert M. Young
14. *The Wohascum County Problem Book,* George T. Gilbert, Mark Krusemeyer, and Loren C. Larson
15. *Lion Hunting and Other Mathematical Pursuits: A Collection of Mathematics, Verse, and Stories by Ralph P. Boas, Jr.,* edited by Gerald L. Alexanderson and Dale H. Mugler
16. *Linear Algebra Problem Book,* Paul R. Halmos
17. *From Erdős to Kiev: Problems of Olympiad Caliber,* Ross Honsberger
18. *Which Way Did the Bicycle Go? ...and Other Intriguing Mathematical Mysteries,* Joseph D. E. Konhauser, Dan Velleman, and Stan Wagon
19. *In Pólya's Footsteps: Miscellaneous Problems and Essays,* Ross Honsberger
20. *Diophantus and Diophantine Equations,* I. G. Bashmakova (Updated by Joseph Silverman and translated by Abe Shenitzer)

MAA Service Center
P. O. Box 91112
Washington, DC 20090-1112
1-800-331-1MAA   fax: 301-206-9789

Preface

This book is a miscellaneous collection of elementary topics that are mostly from algebra, geometry, combinatorics, and number theory. Just as one can hardly fail to pick up something from wandering through an art gallery, there might be some things to be learned from these essays. However, it is not their intention to instruct, but to put on display little gems that are to be found at the elementary level. Their sole objective is to give enjoyment through quality entertainment.

The pace is leisurely and little background is assumed. A college sophomore should be well equipped to have a good time.

The topics are not presented in any particular order. At the end of the essays is a set of exercises (with solutions) which contains some remarkable results. There is also a set of indices to help you locate a particular topic or name in the text.

I cherish the hope that you might be enchanted by these many small wonders of elementary mathematics.

I would like to take this opportunity to thank Professor Dan Velleman and the members of the Dolciani Editorial Board for their warm reception and gentle criticism of the manuscript. The book is much improved because of their dedication and I am deeply grateful to them. It is again a pleasure to extend my warmest thanks to Elaine Pedreira and Beverly Ruedi for their unfailing geniality and technical expertise in seeing the manuscript through publication.

God must love mathematicians—
He's given us so much to enjoy!

Contents

To Don Albers

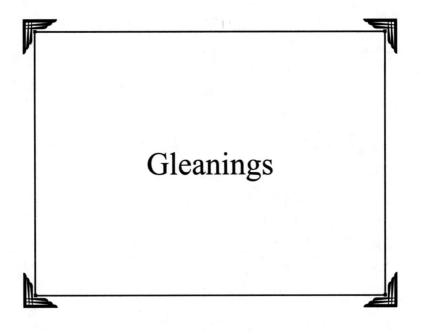

Gleanings

From *Mathematical Miniatures*

Mathematical Miniatures, by Titu Andreescu and Svetoslav Savchev (Anneli Lax New Mathematical Library Series, MAA, 2003), is a goldmine of elementary delights.

1. Six points are given in space such that the lengths of the 15 segments that join them in pairs are all different. Prove that one of these segments is the longest side of one of the triangles it is in and the shortest side of another of them.

   The six points determine $\binom{6}{3} = 20$ scalene triangles. In each of these triangles let the shortest side be colored red. After this is all done, let all the other edges be colored blue (at least the two longest segments will have to be blue).

   Now, any 2-coloring of these edges must produce a monochromatic triangle T (the simple proof of this fact is given in the Appendix at the end of this section). Since every triangle has a shortest side, and hence a red edge, T can only be a red triangle. The longest side in T, then, being red, is also the shortest side in some other triangle (since all the triangles are scalene, it can't be both the longest and shortest side of the same triangle).

2. Now a problem that has a really great solution.

   Cars A, B, C, D travel on the same highway, each moving at its own constant speed. A, B, and C are going in one direction and D in the opposite direction. To begin, A is a distance behind B, who is behind C, and D is far down the highway coming towards them.

   A passed B at 8 A.M. and C at 9 A.M., and was the first to encounter D, whom he met at 10 A.M. D met B at noon, and C at 2 P.M.

   When did B pass C?

In Figure 1, let the position on the highway be plotted along the y-axis against the time along the x-axis. Let the time at the origin O be 8 o'clock and let the point on the highway at the origin be the point where A passes B. Since the speed of a car is constant, the record of its changing positions is given by a straight line in the graph, whose slope is greater the faster the car. Now, it is clear that A goes faster than B and B faster than C. Hence the slopes of their lines decrease from A to B to C. Since D is coming toward A, B, and C, the slope of D's line is negative.

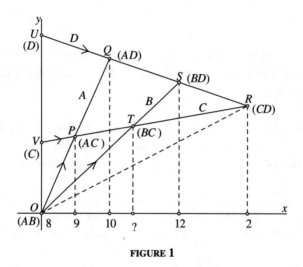

FIGURE 1

At 8 o'clock, then, A and B are together on the highway at the origin O, while C is somewhere ahead of them at V, and D is somewhere down the highway at U. A's speed is indicated by the substantial slope of OP, while C's is given by the much smaller slope of VP, with the result that A catches up to C at 9 o'clock as shown at the point P. A then goes on to meet D at 10 o'clock as indicated at the point Q. Similarly, S and R record D's meetings with B and C at noon and 2 o'clock. The question is: "What time is it at the point T, when B passes C?"

Since the time from 8 to 9 along the x-axis is the same as the time from 9 to 10, the segments OP and PQ along A's line are equal; similarly, $QS = SR$ along D's line. Thus, in $\triangle OQR$, OS and RP are medians, making T the *centroid*! Hence $OT = \frac{2}{3}OS$, and this makes the time at T equal to $\frac{2}{3}$ the time from 8 to 12, that is $[8 + (\frac{2}{3}) \cdot 4]$ o'clock. Hence B passed C at twenty to eleven.

3. (A problem from the 1983 Kurschak Contest)

All the coefficients of

$$f(x) = x^n + a_{n-1}x^{n-1} + \cdots + a_1 x + 1$$

are positive. If the n roots of $f(x) = 0$ are all real, prove that

$$f(2) \geq 3^n.$$

Since all the coefficients are positive, nonnegative values of x make $f(x)$ positive, never zero. Hence the roots of $f(x) = 0$ must all be negative, say $-r_1, -r_2, \ldots, -r_n$. Thus

$$f(x) = (x + r_1)(x + r_2) \cdots (x + r_n),$$

and when multiplied out, the constant term, $r_1 r_2 \cdots r_n$, is equal to 1.

Now, by the A.M.–G.M. inequality, we have

$$2 + r_i = 1 + 1 + r_i \geq 3\sqrt[3]{1 \cdot 1 \cdot r_i}$$
$$= 3\sqrt[3]{r_i}.$$

Hence

$$f(2) = (2 + r_1)(2 + r_2) \cdots (2 + r_n)$$
$$\geq 3\sqrt[3]{r_1} \cdot 3\sqrt[3]{r_2} \cdots 3\sqrt[3]{r_n}$$
$$= 3^n \sqrt[3]{r_1 r_2 \cdots r_n}$$
$$= 3^n \sqrt[3]{1}$$
$$= 3^n.$$

4. **Pompeiu's Theorem.** *If ABC is an equilateral triangle and P a point in its plane, then there exists a triangle with sides of lengths PA, PB, PC.*

(This triangle is nondegenerate unless P lies on the circumcircle of $\triangle ABC$, in which case it is well known that the shorter two of PA, PB, PC add up to the third.)

We consider here only the very simple case of the point P inside the triangle (Figure 2).

Let segments PL, PM, PN be drawn parallel to the sides of the triangle. Then $\angle PLC = \angle B = 60° = \angle C$, making $PLCM$ an isosceles trapezoid. Hence the diagonals PC and LM are equal. Similarly, $PA = MN$ and $PB = LN$, and $\triangle LMN$ has sides of lengths PA, PB, PC.

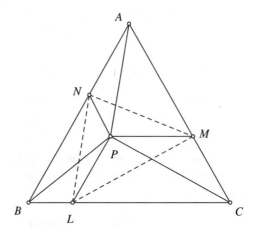

FIGURE 2

5. Asterisks are placed in some cells of an $m \times n$ matrix, where $m < n$, so that there is at least one asterisk in each column. Prove there is an asterisk such that there are more asterisks in its row than in its column.

   Let's go through the rows one at a time replacing each asterisk with the number $\frac{1}{k}$, where k is the number of asterisks in the row under consideration. In this way, the sum of the numbers in a row is either 0, if there is no asterisk in the row, or $k \cdot \frac{1}{k} = 1$, giving a grand total r for the whole matrix, a total which cannot exceed the number of rows: $r \leq m$.

   Now go back to the original matrix and do the same thing for the columns: replace each asterisk with the number $\frac{1}{t}$, where t is the number of asterisks in that column. Then, since each column contains at least one asterisk, the sum of the numbers in a column is always $t \cdot \frac{1}{t} = 1$, giving a grand total of n for the columns. Since $n > m$, and $m \geq r$, it follows that $n > r$.

   Now, both these sums, r and n, contain a term for each asterisk, and hence they can be ordered so that corresponding terms record the values for the same asterisk. In this case, comparing the corresponding entries in the sums, the column sum couldn't exceed the row sum unless the column value for some asterisk were to exceed its corresponding row value: $\frac{1}{t} > \frac{1}{k}$. Thus, for this asterisk we must have $k > t$, which asserts that it is in a row with more asterisks than there are in its column.

6. Finally, let's close this section with a beautiful solution to the difficult last question on the 1988 International Olympiad. Over the years it has acquired a certain notoriety.

If a and b are positive integers such that $a^2 + b^2$ is divisible by $ab + 1$, prove that the complementary divisor,

$$k = \frac{a^2 + b^2}{ab + 1},$$

must be a perfect square.

(a) Proceeding indirectly, let us try to derive a contradiction on the assumption that k is not a perfect square.

Unfortunately, this assumption doesn't do much to suggest a line of attack. It is not inconceivable, however, that at some point in our deliberations the "method of infinite descent" might come to mind and give direction to our efforts as follows.

To start, then, we have that a, b, and k are fixed positive integers and that $k = (a^2+b^2)/(ab+1)$. Since this expression for k is symmetric in a and b, we are at liberty to assign their labels so that $a \geq b$. Now, the crux of the method of infinite descent lies in deducing the existence of a "smaller" pair of positive integers (p, q) such that

$$k = \frac{p^2 + q^2}{pq + 1}, p \geq q \quad \text{and} \quad q < b.$$

Then, in like fashion, the existence of (p, q) would give rise to another pair of positive integers (r, s) such that

$$k = \frac{r^2 + s^2}{rs + 1}, r \geq s \quad \text{and} \quad s < q,$$

which, in turn, would lead to yet another such pair (u, v), with $u \geq v$ and $v < s$, and so on ad infinitum. Thus it would follow that there must exist a strictly decreasing infinite sequence of positive integers,

$$b > q > s > v > \cdots,$$

the impossibility of which would yield the desired conclusion by contradiction. It remains, then, to show there exists a pair (p, q) that is "smaller" than the initial pair (a, b).

(b) The central relation,

$$k = \frac{a^2 + b^2}{ab + 1},$$

is easily put into the form

$$a^2 - kab + b^2 = k,$$

which displays (a, b) as an integer solution of the equation

$$x^2 - kxy + y^2 = k.$$

Now, if $x = 0$ in this equation, then $k = y^2$, and if $y = 0$, then $k = x^2$. Since k is not a square (by assumption), then both components in every solution (x, y) must be nonzero. Moreover, if *integers* x and y were to have opposite signs, then, since k is positive and x and y are nonzero, we would have

$$x^2 - kxy + y^2 = x^2 + k|xy| + y^2$$
$$\geq x^2 + k + y^2 > k, \text{ a contradiction.}$$

Hence in every *integer* solution, x and y must have the same sign.

(c) Recalling that

$$a^2 - kab + b^2 = k,$$

it follows that

$$a^2 - kab + (b^2 - k) = 0,$$

leading to the trivial observation that $x = a$ is a root of the quadratic equation

$$x^2 - kxb + (b^2 - k) = 0.$$

But a quadratic equation has two roots and, in a transport of inspiration, let us turn our attention to this second root. Calling it c, we have

$$c^2 - kcb + (b^2 - k) = 0,$$

making (b, c) a solution of our equation $x^2 - kxy + y^2 = k$. Unraveling the equation, we immediately obtain the major relation

$$k = \frac{b^2 + c^2}{bc + 1}.$$

(d) Now, since a and c are the roots of $x^2 - kxb + (b^2 - k) = 0$, then

(i) their sum $a + c = kb$, and

(ii) their product $ac = b^2 - k$.

From (i) it follows that c is an integer, making (b, c) an integer solution of $x^2 - kxy + y^2 = k$, and since the components of every integer solution have the same sign, c must be positive like b. Thus c is a positive integer.

Recalling that $a \geq b$ and k is positive, (ii) gives

$$ac = b^2 - k < b^2 \leq ab,$$

implying the all-important result that

$$c < b.$$

Thus, in conclusion, we have that if k is not a perfect square, the existence of the positive integers (a, b) such that

$$k = \frac{a^2 + b^2}{ab + 1}, a \geq b,$$

does imply the existence of positive integers b and c such that

$$k = \frac{b^2 + c^2}{bc + 1}, \quad b \geq c \quad \text{and} \quad c < b,$$

completing the proof.

Can you imagine a contestant coming up with what is essentially this argument during the writing of an olympiad? This is just what the young Bulgarian wizard Emanuil Atanasov did, and in recognition of his marvelous achievement he was awarded a special prize.

Appendix

The proof that any 2-coloring of the edges discussed in part 1 of this section must produce a monochromatic triangle T.

Let the six given points be A, B, C, D, E, F, and consider the five edges from A to the points B, C, D, E, F. Since the pigeonhole principle asserts that some three of these edges must be the same color, suppose that AB, AC, and AD are red.

Then any red side of triangle BCD completes an all-red triangle with the edges to A. The only way to avoid this is for BCD itself to be an all-blue triangle, making a monochromatic triangle inevitable.

From *The Contest Problem Book VI*

This book covers the AHSME Competitions (American High School Mathematics Examination) from 1989 to 1994. It is full of wonderful problems, beautifully presented by Leo Schneider (John Carroll University) and appeared in 2000 as volume 40 in the Anneli Lax New Mathematical Library Series.

1 From 1989

(#29). What is the value of the sum

$$S = \sum_{k=0}^{49} (-1)^k \binom{99}{2k} = \binom{99}{0} - \binom{99}{2} + \binom{99}{4} - \cdots - \binom{99}{98}?$$

First we observe that

$$(1+x)^{99} = \binom{99}{0} + \binom{99}{1}x + \binom{99}{2}x^2 + \binom{99}{3}x^3 + \cdots + \binom{99}{99}x^{99}$$

and

$$(1-x)^{99} = \binom{99}{0} - \binom{99}{1}x + \binom{99}{2}x^2 - \binom{99}{3}x^3 + - \cdots - \binom{99}{99}x^{99}.$$

Hence

$$\frac{(1+x)^{99} + (1-x)^{99}}{2} = \binom{99}{0} + \binom{99}{2}x^2 + \binom{99}{4}x^4 + \cdots + \binom{99}{98}x^{98}.$$

This is an identity, and for $x = i = \sqrt{-1}$, we obtain

$$\frac{(1+i)^{99} + (1-i)^{99}}{2} = \binom{99}{0} - \binom{99}{2} + \binom{99}{4} - \cdots - \binom{99}{98} = S.$$

Now, in polar form,

$$1 + i = \sqrt{2}\left(\cos\frac{\pi}{4} + i\sin\frac{\pi}{4}\right).$$

Hence, by the theorem of de Moivre,

$$(1+i)^{99} = \left(\sqrt{2}\right)^{99}\left(\cos\frac{99\pi}{4} + i\sin\frac{99\pi}{4}\right)$$

$$= \left(\sqrt{2}\right)^{99}\left[\cos\left(24\pi + \frac{3\pi}{4}\right) + i\sin\left(24\pi + \frac{3\pi}{4}\right)\right]$$

$$= \left(\sqrt{2}\right)^{99}\left[\cos\frac{3\pi}{4} + i\sin\frac{3\pi}{4}\right].$$

Similarly,

$$(1-i)^{99} = \left(\sqrt{2}\right)^{99}\left[\cos\frac{3\pi}{4} - i\sin\frac{3\pi}{4}\right],$$

and

$$S = \frac{(1+i)^{99} + (1-i)^{99}}{2} = \left(\sqrt{2}\right)^{99}\cos\frac{3\pi}{4}$$

$$= \left(\sqrt{2}\right)^{99}\left(-\frac{1}{\sqrt{2}}\right)$$

$$= -\left(\sqrt{2}\right)^{98}$$

$$= -2^{49}.$$

2 From 1990

(#26). Each of ten girls around a circle chooses a number and tells it to the neighbor on each side. Thus each person gives out one number and receives two numbers. Each girl then announces the average of the two numbers she received. Remarkably, the announced numbers, in order around the circle, were 1, 2, 3, 4, 5, 6, 7, 8, 9, 10.

What was the number chosen by the girl who announced the number 6?

Clearly the sum of the numbers received by a person is twice the average she announces. Since the numbers $1, 2, 3, \ldots, 10$ were announced in order around the circle, the person who announced 5 must have received a total of 10 from her neighbors, namely the girls who announced 4 and 6. Thus if the person who announced 6 chose the number x, the girl who announced 4 must have chosen $10 - x$ (Figure 1).

Working counterclockwise around the circle, consider the girl who announced an average of 3. This requires that she receive a total of 6 and so, in addition to the $10 - x$ she gets from the girl who announced 4, she must have

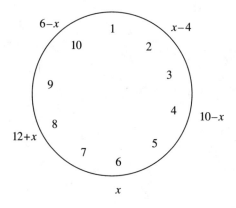

FIGURE 1

received $x - 4$ from the girl who announced 2. Similarly, the person who announced 10 must have chosen the number $6 - x$, and the person who announced 8 must have picked $12 + x$.

Finally, the girl who announced 7 must have received a total of 14, and we have

$$(12 + x) + x = 14, \quad \text{giving } x = 1.$$

From *Problem-Solving Through Problems*

There is so much to enjoy in this wonderful book by Loren Larson (Springer-Verlag, 1983). It has excitement on every page. However, we will consider only one of its problems.

Problem 1.10.5, page 48; from the 1973 Putnam Examination

Let $\{a_1, a_2, \ldots, a_{2n+1}\}$ be a set of integers with the following property P:

P: the removal of any one of the integers leaves a set which can be divided into two subsets of n integers each which have the same sum.

Prove that all the integers must be the same:

$$a_1 = a_2 = \cdots = a_{2n+1}.$$

Since any $2n$ of the integers can be divided into two n-subsets having the same sum, the sum of any $2n$ of the a's must be an even number. It follows that all the integers a_i must have the same parity:

the sum S of all $(2n + 1)$ integers

$= a_i + (a_1 + a_2 + \cdots + a_{i-1} + a_{i+1} + \cdots + a_{2n+1})$

$= a_i + $ (an even number).

Hence each a_i has the same parity as S.

Now suppose, contrary to desire, that the a's are not all equal and that the smallest value among them is a. Thus if a is subtracted from each a_i, at least one of the results $b_i = a_i - a$ will be zero.

Now, these $2n + 1$ integers b_i inherit property P:

Removing any one b_i leaves $2n$ of them which are determined from $2n$ of the a's. Now, these $2n$ a's can be divided into two n-subsets with equal sums T, and subtracting a from each of these a_i yields two n-subsets of b's with the same sum $T - na$.

Accordingly, all the b's must have the same parity, and since one of them is zero, they must all be even.

Since, by assumption, the a's are not all the same, then neither will all the b's be the same. Thus at least one of the b's must be nonzero. Being even, each b_i is divisible by 2, and possibly by higher powers of 2. Of course, zero is divisible by every power of 2, but a nonzero integer is not. Let 2^k be the greatest power of 2 which is a divisor of every b_i. Thus each of $b_1, b_2, \ldots, b_{2n+1}$ is divisible by 2^k, but 2^{k+1}, while possibly dividing some of the b's, does not divide every one of them. Let the complementary divisors be $c_i = b_i/2^k$.

Now, the integers c_i inherit property P from the b's just as the b's got it from the a's:

Removing any one c_i leaves $2n$ of them which come from $2n$ of the b's. Since these b's can be divided into two n-subsets with equal sums R, dividing each of the b's by 2^k converts them into n-subsets of c's each with sum $R/2^k$.

Thus all the c's must have the same parity, and since some b_i is zero, so is the corresponding c_i, implying all the c's are even. That is to say, each c_i is still divisible by 2, implying that each b_i must have been divisible by 2^{k+1} in the first place, not just 2^k as declared above, and the conclusion follows by contradiction.

From *Mathematics Magazine*

1. (Quickie Q879, 1998, page 143, proposed by Jan Mycielski, University of Colorado at Boulder)

   T is a fixed sphere with center O and radius R, and S is a sphere of variable radius r which passes through O (Figure 1). If $r < \frac{1}{2}R$, then S lies completely inside T. In order to avoid this, let $r \geq \frac{1}{2}R$, in which case S and T have a nonempty intersection and T captures a portion of S. Prove the surprising result that the magnitude of the area of the surface of S that lies inside T is the same no matter what the size of S.

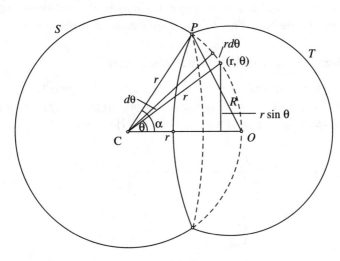

FIGURE 1

   Let C be the center of S and let P be any point on the circle of intersection of S and T. Then $CP = CO = r$, and $PO = R$. Also let $\angle PCO = \alpha$.

   Let C and CO be the origin and axis of a system of polar coordinates. The area A in question is obtained by revolving the circular arc PO about

CO. An element of this arc is $r\,d\theta$ and its distance from the axis is $r\sin\theta$. Hence an element of the area A is given by

$$dA = 2\pi(r\sin\theta)(r\,d\theta),$$

and

$$A = 2\pi r^2 \int_0^\alpha \sin\theta\,d\theta = 2\pi r^2(1 - \cos\alpha).$$

Containing the variables r and α, this hardly looks like a constant.

However, applying the law of cosines to $\triangle PCO$, we obtain

$$R^2 = r^2 + r^2 - 2r \cdot r \cdot \cos\alpha = 2r^2(1 - \cos\alpha),$$

and it follows that $A = 2\pi r^2(1 - \cos\alpha) = \pi R^2$, which is indeed the same for all spheres S.

We might have observed at the outset that, for $r = \frac{1}{2}R$, S is internally tangent to T, in which case T captures the entire surface of S. Thus we could have begun the solution knowing that if the area in question is constant, its value would have to be $4\pi(R/2)^2 = \pi R^2$. Furthermore, we might have noticed that, as r approaches infinity, the surface in question approaches a disk of radius R, which again confirms the answer πR^2.

2. (Problem A-1 from the 1998 Putnam Competition)

Rectangle *HOMF* has sides $HO = 11$ and $OM = 5$ (Figure 2). Now, H is the orthocenter of $\triangle ABC$, O is its circumcenter, M is the foot of the median to BC, and F is the foot of the altitude to BC. How long is BC?

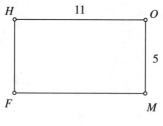

FIGURE 2

The published solution uses analytic geometry and is most appropriate. However, if you happen to recall that the distance down an altitude from a vertex to the orthocenter is twice the distance from the circumcenter to the opposite side, the following nice Euclidean solution is available (this result is proved in part (ii) of the Appendix to this section).

Clearly B and C lie on the line FM which joins two of its feet. Thus AHF is the altitude to BC, and since $OM = 5$, it follows that $AH = 10$ (Figure 3). Therefore, in right triangle AHO, the circumradius

$$R = OA = \sqrt{10^2 + 11^2} = \sqrt{221}.$$

Finally, in right triangle OMC, we have

$$\tfrac{1}{2}BC = MC = \sqrt{R^2 - 5^2} = \sqrt{196} = 14,$$

and $BC = 28$.

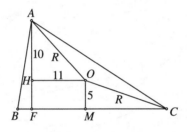

FIGURE 3

3. (From the 1997 U.S.A. Olympiad (1998, page 234))

Outwardly on the sides of $\triangle ABC$ as bases, arbitrary isosceles triangles ABD, BCE, and CAF are drawn to give $\triangle DEF$ (Figure 4).

From A, B, C, perpendiculars are drawn, respectively, to DF, DE, and EF. Prove these perpendiculars are concurrent.

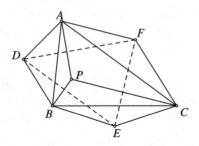

FIGURE 4

We follow the very clever second published solution.

Since $\triangle ABD$ is isosceles, the circle C_1 with center D and radius DA goes through A and B. Similarly, the circle C_2 with center E and radius EB goes through B and C. Thus both these circles go through B, and DE is the line joining their centers. Accordingly, the perpendicular to DE from B is their common chord and their radical axis (Figure 5). (A brief discussion of radical axes is given in part (i) of the Appendix to this section.)

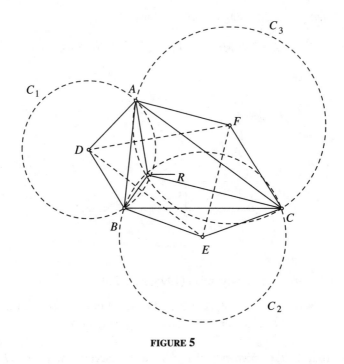

FIGURE 5

Similarly, if C_3 is the circle with center F and radius FC, then the perpendicular from C to EF is the radical axis of C_2 and C_3, and since C_3 clearly goes through A, the perpendicular from A to DF is the radical axis of C_1 and C_3. Hence these perpendiculars are concurrent at the radical center R of C_1, C_2, C_3.

4. (A Nice Observation Concerning Algebraic Numbers (Feb. 2000, page 66, Problem 1566, proposed by Stephen G. Pernice, Morriston, New Jersey; solved by the Con Amore Problem Group, Copenhagen, Denmark))

Let R be a rectangle that is inscribed in a circle C (Figure 6). If α is the ratio of the area of C to the area of R and β is the ratio of the circumference

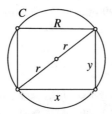

FIGURE 6

of C to the perimeter of R, prove that not both α and β can be algebraic numbers.

Let the radius of C be r and the dimensions of R be x and y. Then

$$\alpha = \frac{\pi r^2}{xy} \quad \text{and} \quad \beta = \frac{2\pi r}{2x + 2y} = \frac{\pi r}{x + y},$$

and we have

$$xy = \frac{\pi r^2}{\alpha} \quad \text{and} \quad x + y = \frac{\pi r}{\beta}.$$

Now, since a diagonal of R is a diameter of C, we have

$$(2r)^2 = x^2 + y^2 = (x + y)^2 - 2xy,$$

giving

$$4r^2 = \frac{\pi^2 r^2}{\beta^2} - \frac{2\pi r^2}{\alpha}.$$

Hence

$$4\alpha\beta^2 = \alpha\pi^2 - 2\beta^2\pi,$$

making π a root of

$$\alpha z^2 - 2\beta^2 z - 4\alpha\beta^2 = 0.$$

Therefore if both α and β are algebraic numbers, so is π, a contradiction. (Although algebraic numbers are defined in terms of polynomial equations with *integer* coefficients, it is known that *algebraic* coefficients also yield algebraic roots.)

5. (Problem A3 from the 1999 Putnam Competition (Feb., 2000, page 74))

Prove that the series expansion

$$\frac{1}{1 - 2x - x^2} = a_0 + a_1 x + \cdots + a_n x^n + \cdots,$$

has the remarkable property that, for every pair of consecutive terms, the sum of the squares of the coefficients occurs as the coefficient of a later term in the series.

The solution to this problem is perfectly straightforward. Let

$$1 - 2x - x^2 = (1 - \alpha x)(1 - \beta x).$$

Then the series is

$$\sum_{n \geq 0} a_n x^n = (1 - \alpha x)^{-1}(1 - \beta x)^{-1}$$

$$= (1 + \alpha x + \cdots + \alpha^n x^n + \cdots)(1 + \beta x + \cdots + \beta^n x^n + \cdots)$$

$$= \cdots + (\alpha^n + \alpha^{n-1}\beta + \alpha^{n-2}\beta^2 + \cdots + \beta^n)x^n + \cdots,$$

giving the coefficient a_n as the sum of $n + 1$ terms of the geometric progression having initial term α^n and common ratio β/α. Hence

$$a_n = \frac{\alpha^n \left[1 - \left(\frac{\beta}{\alpha}\right)^{n+1}\right]}{1 - \frac{\beta}{\alpha}} = \frac{\alpha^{n+1} - \beta^{n+1}}{\alpha - \beta}.$$

Now, since $1 - 2x - x^2 = (1 - \alpha x)(1 - \beta x)$, we have $\alpha\beta = -1$, from which $\frac{1}{\alpha} = -\beta$ and $\frac{1}{\beta} = -\alpha$. Hence

$$\alpha + \frac{1}{\alpha} = \alpha - \beta \quad \text{and} \quad \beta + \frac{1}{\beta} = \beta - \alpha = -(\alpha - \beta).$$

Finally, then,

$$a_n^2 + a_{n+1}^2 = \left(\frac{\alpha^{n+1} - \beta^{n+1}}{\alpha - \beta}\right)^2 + \left(\frac{\alpha^{n+2} - \beta^{n+2}}{\alpha - \beta}\right)^2$$

$$= \frac{1}{(\alpha - \beta)^2}\left[\alpha^{2n+2} - 2(\alpha\beta)^{n+1} + \beta^{2n+2} + \alpha^{2n+4}\right.$$

$$\left. -2(\alpha\beta)^{n+2} + \beta^{2n+4}\right]$$

$$= \frac{1}{(\alpha - \beta)^2}\left[\alpha^{2n+3}\left(\alpha + \frac{1}{\alpha}\right) + \beta^{2n+3}\left(\beta + \frac{1}{\beta}\right)\right]$$

(recall $\alpha\beta = -1$)

$$= \frac{1}{(\alpha - \beta)^2} \left[\alpha^{2n+3}(\alpha - \beta) - \beta^{2n+3}(\alpha - \beta) \right]$$

$$= \frac{\alpha^{2n+3} - \beta^{2n+3}}{\alpha - \beta} = a_{2n+2}.$$

It is easy to check the first few cases. From the definition,

$$1 = (1 - 2x - x^2)(a_0 + a_1 x + a_2 x^2 + \cdots + a_n x^n + \cdots)$$
$$= a_0 + (a_1 - 2a_0)x + \cdots + (a_n - 2a_{n-1} - a_{n-2})x^n + \cdots,$$

in which $a_0 = 1$ and all the other coefficients are zero. Hence

$$a_0 = 1, \quad a_1 = 2, \quad \text{and in general, } a_n = 2a_{n-1} + a_{n-2}.$$

Thus the series begins

$$1 + 2x + 5x^2 + 12x^3 + 29x^4 + 70x^5 + 169x^6 + 408x^7 + 985x^8 + \cdots,$$

and we have

$$a_0{}^2 + a_1{}^2 = 1 + 4 = 5 = a_2,$$
$$a_1{}^2 + a_2{}^2 = 4 + 25 = 29 = a_4,$$
$$a_2{}^2 + a_3{}^2 = 25 + 144 = 169 = a_6,$$
$$a_3{}^2 + a_4{}^2 = 144 + 841 = 985 = a_8.$$

We observe that the solution makes no use of the 2 in the denominator of the given expression and therefore, for any integer c, the expansion of

$$\frac{1}{1 + cx - x^2}$$

enjoys the intriguing property of this section.

6. (Problem A4 from the 1998 Putnam Competition (Feb. 1999, page 74))

Let the sequence $\{A_n\}$ be constructed with $A_1 = 0$, $A_2 = 1$ and, for $n \geq 3$, let A_n be the integer formed by concatenating the digits of A_{n-1} and A_{n-2}, in that order. Thus the sequence begins

$$\{0, 1, 10, 101, 10110, 10110101, \ldots\}.$$

Which A_n are divisible by 11?

We follow the published solution.

Every Putnam competitor knows the test for divisibility by 11: an integer $n = abcd \ldots$, in decimal notation, is divisible by 11 if and only if the alternating sum of its digits,

$$a - b + c - d + - \cdots,$$

is divisible by 11.

Let r_n and d_n, respectively, denote the number of digits and the alternating sum of the digits of A_n. From the concatenation

$$A_n = (\ldots\ldots)\,(\ldots\ldots),$$
$$\underset{A_{n-1}}{\qquad} \underset{A_{n-2}}{\qquad}$$

it follows that

$$d_n = d_{n-1} + d_{n-2} \quad \text{when } r_{n-1} \text{ is even}$$

and

$$d_n = d_{n-1} - d_{n-2} \quad \text{when } r_{n-1} \text{ is odd.}$$

In every case, then,

$$d_n = d_{n-1} + (-1)^{r_{n-1}} d_{n-2}.$$

Whether d_{n-2} is added or subtracted depends only on the parity of r_{n-1}. Now, it is evident from the concatenation that

$$r_n = r_{n-1} + r_{n-2},$$

and since r_1 and r_2 are obviously 1, the sequence $\{r_n\}$ begins

$$\{r_n\} = \{1, 1, 2, 3, 5, 8, 13, 21, \ldots\}.$$

We can't help brightening up upon recognizing this to be the Fibonacci sequence, but unfortunately the observation doesn't do anything to advance our cause. What is important, however, is that, because $r_n = r_{n-1} + r_{n-2}$, the parities of r_n cycle with a period of length 3,

$$(\text{odd, odd, even}), (\text{odd, odd, even}), \ldots,$$

making $(-1)^{r_{n-1}}$ repeatedly cycle through the values $(-1, -1, +1)$. Beginning with the odd integer r_4, then, we have

$$d_5 = d_4 - d_3,$$
$$d_6 = d_5 - d_4,$$

and

$$d_7 = d_6 + d_5.$$

From $\{An\} = \{0, 1, 10, 101, 10110, 10110101, \ldots\}$, we get the initial values $d_1 = 0, d_2 = 1, d_3 = 1$, and $d_4 = 2$, and carrying on with the recursion, we can easily calculate as many values of d_n as we have the patience for. Thus

$$d_3 = 1,$$
$$d_4 = 2,$$
$$d_5 = d_4 - d_3 = 2 - 1 = 1,$$
$$d_6 = d_5 - d_4 = 1 - 2 = -1,$$
$$d_7 = d_6 + d_5 = -1 + 1 = 0,$$
$$d_8 = d_7 - d_6 = 0 - (-1) = 1,$$
$$d_9 = d_8 - d_7 = 1 - 0 = 1,$$
$$d_{10} = d_9 + d_8 = 1 + 1 = 2,$$
$$d_{11} = d_{10} - d_9 = 2 - 1 = 1,$$
$$d_{12} = d_{11} - d_{10} = 1 - 2 = -1,$$
$$d_{13} = d_{12} + d_{11} = -1 + 1 = 0,$$
$$- - - - - - - - - - - - - - - - \, .$$

In calculating d_{11}, we observe that the values of d_{10} and d_9 are respectively equal to the values of d_4 and d_3 that go into the calculation of d_5. Also, both these calculations are at the same point in the $(-, -, +)$ cycle of signs that are attached to d_{n-2}, so that d_{11} and d_5 are both given by the same expression. Moreover, the calculations of d_{12} and d_6 continue to be at the same point in the $(-, -, +)$ cycle of signs. Thus d_{12} and d_6 are also the same, and it is clear that the conditions for

$$d_{11+k} = d_{5+k}$$

are perpetuated. Therefore $\{d_n\}$ repeats with a period of length 6, beginning at least at d_5. In fact, $\{d_n\}$ repeats from the beginning, and we conclude that the only values taken by d_n are those in its period, namely

$$0, 1, 2, \quad \text{and} \quad -1.$$

Thus the only times 11 divides d_n is when $d_n = 0$, which happens every sixth term, beginning with d_1. Hence the A_n which are divisible by 11 are

$$\{A_1, A_7, A_{13}, A_{19}, \ldots, A_{6k+1}, \ldots\}.$$

7. (A Surprising Property of a Circle (Problem A2 of the 1998 Putnam Competition))

Let C be a circle with center at the origin O of a system of rectangular coordinates, and let MON be the quarter-circle of C in the first quadrant (Figure 7). Let PQ be an arc of C, of *fixed* length, that lies in the arc MN.

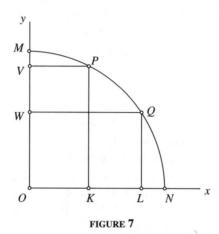

Let A be the area of the region in the quadrant that lies below PQ and B the area in the quadrant to the left of PQ, that is,

$$A = \text{area } PKLQ \quad \text{and} \quad B = \text{area } PVWQ.$$

Prove that $A + B$ is always the same no matter where PQ might occur on arc MN.

We follow the published solution.

Let PK and WQ intersect at U (Figure 8), and let R denote the area of the rectangle $UKLQ$, S the area of rectangle $PVWU$, and T the area of the curved triangle PUQ.

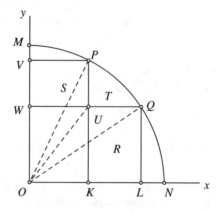

Then

$$A = R + T \quad \text{and} \quad B = S + T, \quad \text{and} \quad A + B = R + S + 2T.$$

Now, a rectangle has twice the area of a triangle having the same base and lying between the same parallels, and so

$$R = 2\triangle UOQ \quad \text{and} \quad S = 2\triangle POU,$$

implying

$$A + B = 2(\triangle UOQ + \triangle POU + T)$$
$$= 2 \cdot \text{sector } POQ,$$

which, having an arc of fixed length, is the same wherever it occurs on *MN*.

8 Schoch 15

This section is based on the captivating note *Those Ubiquitous Archimedean Circles* by Clayton Dodge, Thomas Schoch, Peter Y. Woo, and Paul Yiu (June, 1999, pages 202–213).

1. The arbelos, or shoemaker's knife, is bounded by three tangent semicircles (O), (O_1), (O_2), of radii r, r_1, r_2, as shown in Figure 9.

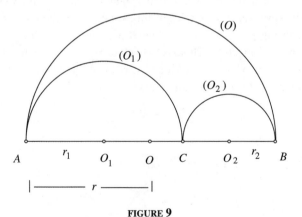

FIGURE 9

The arbelos has a long history, going back to the ancient Greeks, and it has led to the discovery of many fascinating properties. Prominent among them are the "twins of Archimedes": let the common tangent *CD* be drawn

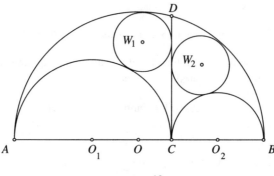

FIGURE 10

to (O_1) and (O_2) (Figure 10). Then the circles (W_1) and (W_2) inscribed in the regions ACD and BCD are the same size.

As we shall see shortly, this is established by easy applications of the theorem of Pythagoras.

In 1974, the renowned Los Angeles dentist Leon Bankoff pointed out that the twins are actually just two of a set of triplets:

> if the inscribed circle (O_3) of the arbelos touches (O_1) and (O_2) at U and V, then the circle (W_3) through U, V, and C is the same size as the twins (Figure 11).

Well, that started the ball rolling, and before long, people were making wonderful discoveries of circles of this same size in the most unexpected and interesting places in this configuration. One of the most attractive of these gems was given by Thomas Schoch of Essen, Germany, in 1979. He found a number of these circles and it is his circle (W_{15}) (so labeled by Dodge et al.) that is our interest in this section.

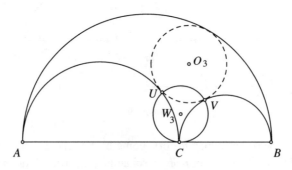

FIGURE 11

Schoch's (W_{15})

The inscribed circle in the region enclosed by (O) and arcs CS and CT having centers at A and B and radii AC and BC, respectively, is yet another Archimedean circle (Figure 12).

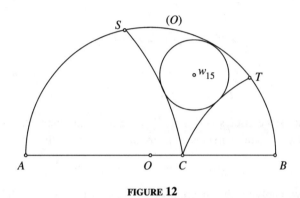

FIGURE 12

2. In order to show that (W_{15}) is the same size as (W_1) and (W_2) we first need to determine their common radius ρ.

   Let the perpendicular from W_1 meet AB at Q (Figure 13), making $QC = \rho$. Now, clearly $AB = 2r = 2r_1 + 2r_2$, and so

$$r = r_1 + r_2.$$

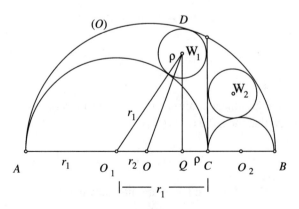

FIGURE 13

Hence

$$O_1 O = AO - AO_1 = r - r_1 = r_2,$$

from which

$$OC = r_1 - r_2.$$

Thus,

$$O_1 Q = O_1 C - QC = r_1 - \rho,$$

and

$$OQ = O_1 Q - r_2 = r_1 - r_2 - \rho.$$

Also, since circles (O) and (W_1) are internally tangent, the distance OW_1 between their centers is equal to the difference of their radii, giving

$$OW_1 = r - \rho = r_1 + r_2 - \rho.$$

Now, applying the theorem of Pythagoras to triangles $O_1 W_1 Q$ and $OW_1 Q$, we have

$$W_1 Q^2 = O_1 W_1{}^2 - O_1 Q^2 = OW_1{}^2 - OQ^2,$$

giving

$$(r_1 + \rho)^2 - (r_1 - \rho)^2 = (r_1 + r_2 - \rho)^2 - (r_1 - r_2 - \rho)^2.$$

Factoring as the difference of squares, we have

$$2r_1 \cdot 2\rho = (2r_1 - 2\rho) \cdot 2r_2,$$
$$r_1 \rho = r_1 r_2 - r_2 \rho,$$

giving

$$\rho = \frac{r_1 r_2}{r_1 + r_2}.$$

A similar argument confirms that (W_2) also has radius ρ. It remains, then, to show that the radius k of (W_{15}) is

$$\frac{r_1 r_2}{r_1 + r_2}.$$

3. Observe that, since (O) and (W_{15}) are internally tangent,

$$OW_{15} = r - k = r_1 + r_2 - k \qquad \text{(Figure 14)}.$$

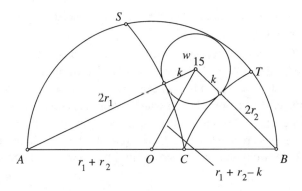

FIGURE 14

Happily, the desired result follows immediately from the well known theorem that the sum of the squares of two sides of a triangle is equal to twice the square of the median to the third side plus twice the square of one-half the third side (this follows immediately from two applications of the law of cosines). Observing that OW_{15} is the median to AB in $\triangle ABW_{15}$, we have

$$AW_{15}^2 + W_{15}B^2 = 2W_{15}O^2 + 2AO^2,$$
$$(2r_1 + k)^2 + (2r_2 + k)^2 = 2(r_1 + r_2 - k)^2 + 2(r_1 + r_2)^2,$$

i.e.,

$$4r_1^2 + 4r_1k + k^2 + 4r_2^2 + 4r_2k + k^2$$
$$= 2(r_1^2 + r_2^2 + k^2 + 2r_1r_2 - 2r_1k - 2r_2k) + 2(r_1^2 + 2r_1r_2 + r_2^2),$$

which easily reduces to $8r_1k + 8r_2k = 8r_1r_2$, and we have

$$k = \frac{r_1r_2}{r_1 + r_2}.$$

9. (From the 1998 U.S.A. Olympiad (June, 2000, page 248))

A computer screen shows a 98×98 chessboard, colored in the usual way. One can select with a mouse any rectangle with sides on the lines of the chessboard and click the mouse button: as a result, the colors in the selected rectangle switch (black becomes white and white becomes black). Determine the minimum number of mouse-clicks needed to make the chessboard all one color.

I was very pleased to see this problem, not just because it is an excellent problem, but because it gives me a chance to slip in the following old favorite: what is the minimum number of bishops needed to control all the squares on a regular chessboard?

(a) The Bishop Problem. A regular chessboard is an 8×8 square and therefore has $2(8 + 6) = 28$ squares in its border, 14 black and 14 white. It is not difficult to see that a bishop can control not more than 4 squares on the perimeter. Therefore at least 4 white bishops are required, for 3 of them can cover at most 12 of the 14 white squares around the outside. Similarly, 4 black bishops are needed, and since a line of 8 bishops across the board in one of the two middle ranks does control the entire board, the minimum is 8.

(b) The 98×98 Board. Again the key is to look at the squares around the outside of the board. Since a 98×98 board tends to boggle the mind with its 9604 squares, it will probably be easier on us to consider the general case. In the perimeter of an $n \times n$ board there are

$$2[n + (n - 2)] = 4n - 4 = 4(n - 1) \text{ squares.}$$

Let α denote the number of times two consecutive squares in the perimeter are of different colors. At the beginning, then, $\alpha = 4(n - 1)$:

since the colors alternate around the border, every pair of consecutive squares is mismatched (squares (1 and 2), (2 and 3), (3 and 4), ..., $(4(n-1)$ and 1)).

But after all the mouse-clicking is done, α is down to zero. Now, it is not difficult to see that no mouse click can reduce α by more than 4:

(i) If the selected rectangle doesn't intersect the border, α remains unchanged.

(ii) A rectangle that touches a single edge of the border affects only the two pairs at the edges of the rectangle. Of course, the switching of colors in the rectangle could create a mismatched pair and increase the value of α. Since it affects only two pairs in the perimeter, however, it cannot increase or decrease α by more than 2 (Figure 15). Similarly, a rectangle that goes right across the board to touch the opposite side could alter α only by as much as 4.

(iii) If the rectangle contains a corner, again we have that α cannot be changed by more than 2 (Figure 16).

Thus, in order to reduce α from $4(n - 1)$ to zero in $n - 1$ steps, each step would be obliged to lower it by the maximum amount 4. Since it is

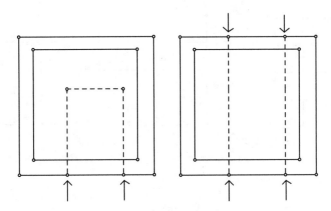

FIGURE 15

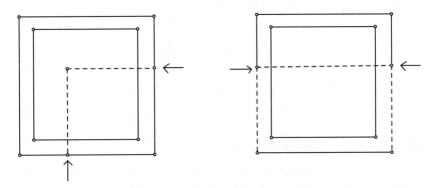

FIGURE 16

impossible to do better than this, $n-1$ mouse-clicks are certainly necessary, and if you have to deal with a corner, thus dropping α by at most 2 at some step, more than $n-1$ steps will be needed. This is the situation when n is even, for then adjacent corners are opposite in color and require attention. Thus, for n even, at least n mouse-clicks are necessary.

However, it is easy to see that n clicks suffice when n is even: simply operate, one line at a time, on the even-numbered columns, and then on the even-numbered rows, as illustrated in Figure 17.

Hence the minimum for a 98×98 board is 98.

When n is an odd number $2k+1$, however, all the corners are the same color and are avoided by the same procedure (Figure 18): doing the k even-numbered columns and then the k even-numbered rows gives a monochromatic board in a total of $2k = n-1$ clicks.

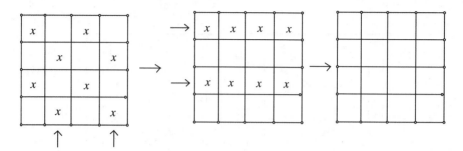

FIGURE 17

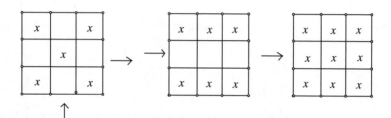

FIGURE 18

10 A Note on Odd Perfect Numbers

(Based on the excellent note "The Abundancy Ratio, A Measure of Perfection" by Paul A. Weiner, Saint Mary's University of Minnesota, Winona, Minnesota; Oct. 2000, pages 307–310)

Recall that the arithmetic function $\sigma(n)$ is the sum of the positive divisors of the positive integer n, including 1 and n, and that n is a perfect number if and only if $\sigma(n) = 2n$.

Fewer than 40 perfect numbers are known, all of them even. It is an open question whether an odd perfect number exists, but it is known that it would have to be a number of more than 300 digits. In this section we shall establish the delightful result that if

$$\frac{\sigma(n)}{n} = \frac{5}{3}$$

for a positive integer n, then $5n$ is an odd perfect number.

The proof is perfectly straightforward but we do need to review a few basic properties of $\sigma(n)$.

Preliminaries

(i) If the prime decomposition of n is

$$n = p_1^{a_1} p_2^{a_2} \cdots p_k^{a_k},$$

then the divisors of n are the numbers

$$d = p_1^{b_1} p_2^{b_2} \cdots p_k^{b_k}, \quad \text{where } 0 \le b_i \le a_i.$$

Thus an expression for $\sigma(n)$ is

$$\sigma(n) = \prod_{i=1}^{k} (1 + p_i + p_i^2 + \cdots + p_i^{a_i})$$

$$= (1 + p_1 + p_1^2 + \cdots + p_1^{a_1})(1 + p_2 + p_2^2 + \cdots + p_2^{a_2})$$

$$\cdots (1 + p_k + p_k^2 + \cdots + p_k^{a_k});$$

when this product is expanded, only the divisors d occur in the sum and none of them is left out.

(ii) (a) Clearly the factor $(1 + p_i + p_i^2 + \cdots + p_i^{a_i})$ contains $a_i + 1$ terms. Now, if n is odd, all its prime divisors p_i are odd, and if it is also known that $\sigma(n)$ is odd, then each of its factors $(1 + p_i + p_i^2 + \cdots + p_i^{a_i})$ is odd. Then this factor would be the sum of $a_i + 1$ odd numbers, implying $a_i + 1$ is yet another odd number. Thus each a_i must be even and we conclude that, when n and $\sigma(n)$ are both odd, n is a square.

(b) It is almost immediate that $\sigma(n)$ is multiplicative, that is, that

$$\sigma(mn) = \sigma(m)\sigma(n) \quad \text{when } m \text{ and } n \text{ are relatively prime.}$$

If the prime decompositions of n and m are

$$n = \prod_{i=1}^{k} p_i^{a_i} \quad \text{and} \quad m = \prod_{i=1}^{t} q_i^{b_i},$$

where no primes p, q are equal, then the prime decomposition of mn is

$$mn = p_1^{a_1} p_2^{a_2} \cdots p_k^{a_k} q_1^{b_1} q_2^{b_2} \cdots q_t^{b_t}, \quad \text{and}$$

$$\sigma(mn) = (1 + p_1 + \cdots + p_1^{a_1})$$

$$\cdots (1 + p_k + \cdots + p_k^{a_k})(1 + q_1 + \cdots + q_1^{b_1})$$

$$\cdots (1 + q_t + \cdots + q_t^{b_t})$$

$$= \sigma(n)\sigma(m).$$

(iii) Suppose the divisors of n are

$$d_1 = 1 < d_2 < d_3 < \cdots < d_r = n.$$

Then the numbers

$$\frac{n}{d_1}, \frac{n}{d_2}, \ldots, \frac{n}{d_r}$$

constitute the same set of numbers in reverse order. Thus both sets have the same sum, and we have

$$\sigma(n) = \frac{n}{d_1} + \frac{n}{d_2} + \cdots + \frac{n}{d_r}$$

$$= n\left(\frac{1}{d_1} + \frac{1}{d_2} + \cdots + \frac{1}{d_r}\right).$$

Thus $\sigma(n)/n$ is just the sum of the reciprocals of the divisors of n.

Corollary. *If m divides n, then a divisor of m is a divisor of n, and the reciprocal of a divisor of m is the reciprocal of a divisor of n. Hence the sum of the reciprocals of the divisors of n is at least as great as the sum of the reciprocals of the divisors of m:*

$$\textit{if } m \textit{ divides } n, \textit{ then } \frac{\sigma(n)}{n} \geq \frac{\sigma(m)}{m}.$$

(iv) Finally, we observe that, since $\sigma(n) = 2n$ is the definition of a perfect number,

$$\frac{\sigma(n)}{n} = 2$$

if and only if n is a perfect number.

Now to the easy proof.

Proof: We will show that if

$$\frac{\sigma(n)}{n} = \frac{5}{3},$$

then $5n$ is odd and

$$\frac{\sigma(5n)}{5n} = 2.$$

From

$$\frac{\sigma(n)}{n} = \frac{5}{3},$$

we have $3\sigma(n) = 5n$, implying 3 divides n. Now, if n were even, then 6 would divide n, and we would have

$$\frac{\sigma(n)}{n} \geq \frac{\sigma(6)}{6} = \frac{12}{6} > \frac{5}{3} = \frac{\sigma(n)}{n},$$

a contradiction. Hence n must be odd, in which case $5n$ is also odd.

Then, from $3\sigma(n) = 5n$, it follows that $\sigma(n)$ must be odd, too, and, as we saw above, this fact, combined with n odd, implies that n is a square. Therefore, since the prime 3 divides n, so does 3^2.

Now let's see whether 5 divides n. If 5 were to divide n, then $3^2 \cdot 5 = 45$ would divide n, and we would have

$$\frac{\sigma(n)}{n} \geq \frac{\sigma(45)}{45} = \frac{1 + 3 + 5 + 9 + 15 + 45}{45} = \frac{78}{45} = \frac{26}{15} > \frac{5}{3} = \frac{\sigma(n)}{n},$$

a contradiction. Hence 5 does not divide n and we conclude that 5 and n are relatively prime.

From the multiplicity of $\sigma(n)$, then, we have

$$\frac{\sigma(5n)}{5n} = \frac{\sigma(5)\sigma(n)}{5n} = \frac{6 \cdot \sigma(n)}{5n} = \frac{6}{5} \cdot \frac{\sigma(n)}{n} = \frac{6}{5} \cdot \frac{5}{3} = 2,$$

as desired.

11 An Elementary Inequality

(From the most enjoyable note *Boxes for Isoperimetric Triangles*, by John Wetzel, University of Illinois at Urbana-Champaign, Oct. 2000, page 315–319)

Here's a nice little problem. If triangle ABC has perimeter 2, prove that not all its altitudes can exceed $1/\sqrt{3}$.

Let $BC = a$ be a longest side of $\triangle ABC$. Then the altitude h_a to BC is a shortest altitude. Since a longest side must be at least $\frac{1}{3}$ the perimeter, we have $a \geq \frac{2}{3}$, from which $\frac{3}{2}a \geq 1$. Observing that the semiperimeter of $\triangle ABC$ is $s = 1$ and that its area is $\Delta = \frac{1}{2}ah_a$, we have

$$h_a = 1 \cdot h_a \leq \left(\frac{3}{2}a\right)h_a = 3\Delta = 3\sqrt{1(1-a)(1-b)(1-c)}.\dots \quad (1)$$

by Heron's formula. (Two proofs of Heron's formula are given in part 7 of section 5.)

Now, applying the arithmetic mean-geometric mean inequality to the numbers $1 - a, 1 - b, 1 - c$, we obtain

$$\frac{1 - a + 1 - b + 1 - c}{3} \geq [(1 - a)(1 - b)(1 - c)]^{1/3}.$$

Since the perimeter $a + b + c = 2$, this yields

$$\tfrac{1}{3} \geq [(1 - a)(1 - b)(1 - c)]^{1/3},$$

from which

$$\Delta = \sqrt{(1 - a)(1 - b)(1 - c)} \leq \left(\tfrac{1}{3}\right)^{3/2} = \tfrac{1}{3\sqrt{3}},$$

giving

$$3\Delta \leq \tfrac{1}{\sqrt{3}}.$$

Hence, from (1) above,

$$h_a \leq 3\Delta \leq \tfrac{1}{\sqrt{3}},$$

as desired.

12 A Nice Problem in Probability

(Problem 1582, part (a), proposed by the Western Maryland College Problems Group, Westminster, Maryland; solved by the SUNY Oswego Problems Group, Oswego, New York; Oct. 2000, page 326)

Teams A and B play a series of games. Each game has three possible outcomes: A wins, B wins, or they tie. The probability that A wins is p, the probability that B wins is q, and consequently the probability of a tie is $r = 1 - p - q$. The series ends when one team has won two more games than the other, that team being declared the winner of the series.

What is the probability that A wins the series?

The sticky point, of course, is how to deal with ties. If we let A denote a game that was won by A, B a game that was won by B, and t a game that ended in a tie, then a record of the series is given by some sequence of A's, B's, and t's like

$$t \, A \, t \, t \, B \, A \, t \, B \, t \, t \, t \, B \, t \, t \, A \, \ldots.$$

Let this string be partitioned by taking with each A, and with each B, all the t's that immediately precede it:

$$(t A)(t t B)(A)(t B)(t t t B)(t t A) \ldots.$$

Thus each group represents a round of games that results in a team being credited with an advance in the standings of the series.

Now, a round can contain any number of ties, and so a round that is won by A belongs to the sequence

$$A, \ tA, \ ttA, \ tttA, \ ttttA, \ \ldots,$$

the members of which occur with the respective probabilities

$$p, rp, r^2p, r^3p, r^4p, \ldots$$

Therefore the probability p' of A winning a round of play is

$$p' = p + rp + r^2p + \cdots = \frac{p}{1-r} = \frac{p}{1-(1-p-q)} = \frac{p}{p+q}.$$

Similarly, the probability of B winning a round is

$$q' = \frac{q}{p+q}.$$

We observe that $p' + q' = 1$.

Summarizing, then, the probability of team A coming out the winner is the probability of A winning a series of rounds in which its probability of winning a round is p' and its probability of losing a round is q'.

Now, any series that is won by a lead of two games, that is to say, two rounds, must last an even number of rounds: some n are won by the loser and $n + 2$ by the winner. Clearly there are only two states at the end of an intermediate round—either the teams are tied in the standings, or one team must be ahead by a single round. Thus, after an *even* number of intermediate rounds, the teams have to be tied, for a lead of one round is attainable only after an odd number of rounds.

Thus, if the series lasts a total of $2n + 2$ rounds, they must be tied at the end of 2 rounds, then again after 4 rounds, and also after 6 rounds, $\ldots$, and finally tied at $2n$ rounds, after which the winner takes the last two rounds. That is to say, they must have split each of the n pairs of rounds $(1, 2), (3, 4), \ldots,$ $(2n - 1, 2n)$. Now, since A might win the first round of a pair and lose the second, or vice versa, the probability of splitting a pair of rounds is

$$p'q' + q'p' = 2p'q'.$$

Thus the probability of A winning the series in $2n + 2$ rounds is

$$(2p'q')^n \cdot (p')^2,$$

and its probability of winning the series at some juncture is the sum of the geometric series

$$\sum_{n=0}^{\infty}(2p'q')^n(p')^2 = (p')^2 + (p')^2(2p'q') + (p')^2(2p'q')^2 + \cdots$$

$$= \frac{(p')^2}{1-2p'q'}$$

$$= \frac{\left(\frac{p}{p+q}\right)^2}{1-\frac{2pq}{(p+q)^2}}$$

$$= \frac{p^2}{p^2+q^2}.$$

13 Spanning Trees in a $K_{2,n}$

Let the $m+n$ vertices of a graph G be partitioned into a set X of m vertices and a set Y of n vertices. If no edge joins two vertices in X or two vertices in Y, but every vertex in X is joined to every vertex in Y, then G is said to be the complete bipartite graph $K_{m,n}$.

A *subgraph* of a graph is any subset of its vertices and edges in which each edge has both its endpoints.

A *spanning subgraph* of G is one that contains all the vertices of G.

A *path* is an alternating sequence $v_0e_1v_1e_2v_2\ldots$ of vertices v_i and edges e_i in which, for all $i \geq 1$, v_{i-1} and v_i are the endpoints of e_i, and in which no vertex is repeated.

A graph is *connected* if there is a path between every pair of its vertices.

A *cycle* C is a connected subgraph in which each vertex of C is the endpoint of exactly two edges. Alternatively, a cycle is a path to which an edge has been added to join its initial and final vertices.

Finally, a *tree* is a connected graph that doesn't have any cycles.

Thus a spanning tree of a graph is a subgraph that is

(i) connected,
(ii) contains all the vertices, and
(iii) doesn't have any cycles.

How many spanning trees T are there in a $K_{2,n}$ (Figure 19)?

Let the parts of the $K_{2,n}$ be X and Y, and let X contain just the two vertices A and B; also, let T be a spanning tree. Since T is a connected spanning

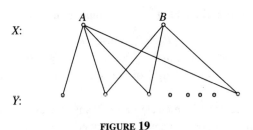

FIGURE 19

subgraph, each vertex of the graph must be the endpoint of at least one edge in T. Now, from a vertex u in Y, there are only the two edges uA and uB available, and so, in any T, each vertex u must be joined to at least one of A and B. In T, let L be the set of vertices that are joined to A, and let M be the set of vertices that are joined to B (Figure 20).

If L and M have no vertex in common, T would fail to be connected (Figure 20(a)), and if L and M were to overlap in as many as two vertices u and v, then T would contain a forbidden cycle $AuBv$ (Figure 20(b)). It follows that L and M must have exactly one vertex u in common, and that each of the other $n-1$ vertices in Y is joined either to A or to B, but not both (Figure 20(c)).

Clearly, there are n ways to single out a vertex u in Y and two ways to link each of the remaining $n-1$ vertices in Y to either A or B, for a total of $n \cdot 2^{n-1}$ possible spanning trees T in $aK_{2,n}$.

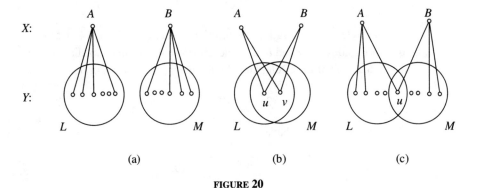

FIGURE 20

For six other neat ways to derive this result, see the excellent note "Spanning Trees: Let Me Count the Ways" by Douglas Shier of Clemson University (Dec., 2000, pages 376–381).

14. (Problem A3 from the 2000 Putnam Competition (Feb. 2001, page 77))

The octagon $P_1 P_2 P_3 P_4 P_5 P_6 P_7 P_8$ is inscribed in a circle, with the vertices around the circumference in the given order. $P_1 P_3 P_5 P_7$ is a square of area 5 and $P_2 P_4 P_6 P_8$ is a rectangle of area 4. What is the maximum possible area of the octagon?

Let the center of the circle be O. Many circles have inscribed *rectangles* of area 5 but only one has an inscribed *square* of area 5. The side of the square is obviously $\sqrt{5}$, and it is clearly the hypotenuse of the isosceles right triangle $P_1 O P_7$ whose legs are the radius r of the circle (Figure 21). Hence

$$2r^2 = 5 \quad \text{and} \quad r = \frac{\sqrt{5}}{\sqrt{2}}.$$

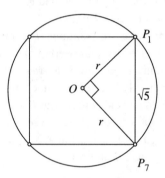

In a circle of given radius an inscribed rectangle of area 4 is also unique. If its sides are a and b, then clearly $ab = 4$, and since its diagonal is also a diameter of the circle, then, by the theorem of Pythagoras,

$$a^2 + b^2 = (2r)^2 = 10, \quad \text{since } r = \frac{\sqrt{5}}{\sqrt{2}}.$$

Solving with $ab = 4$, we have

$$a^2 + \frac{16}{a^2} = 10,$$
$$a^4 - 10a^2 + 16 = 0.$$
$$(a^2 - 2)(a^2 - 8) = 0.$$

Thus $a^2 = 2$ or 8, and $b^2 = 10 - a^2 = 8$ or 2. The sides of the rectangle are therefore $\sqrt{2}$ and $2\sqrt{2}$.

The problem, of course, is to figure out how the square and the rectangle should overlap for maximum area.

Consider just the circle and the rectangle. The square must have a vertex in each minor arc that is cut off by a side of the rectangle, and each of these vertices determines a triangle with the nearest side of the rectangle, like $\triangle P_3 P_4 P_2$ in Figure 22. In fact, the octagon consists precisely of the rectangle and its four bordering triangles (Figure 23). The idea, then, is to make the total area of these four *triangles* as large as possible.

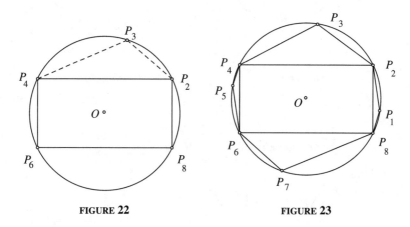

FIGURE 22   FIGURE 23

Consider $\triangle P_2 P_3 P_4$. Wherever P_3 might occur on its arc, the base of the triangle is always $P_2 P_4$, and so this triangle is greatest when the altitude to $P_2 P_4$ is greatest. Clearly this occurs for P_3 at the midpoint of the arc. But the beauty of this is that, because of the symmetry of the rectangle and the square, putting the vertex P_3 at the midpoint of the arc $P_2 P_4$, automatically puts the other vertices of the square at the midpoints of their arcs, thus simultaneously making each of the four triangles as large as possible and thereby making the greatest possible octagon (Figure 24). It remains only to perform the easy calculations.

The diameter $P_4 O P_8$ bisects the the rectangle and the octagon, and we have (Figure 25)

$$\text{the area of the octagon} = 2[\triangle P_2 P_3 P_4 + \triangle P_2 P_4 P_8 + \triangle P_2 P_8 P_1].$$

Recalling that $P_2 P_8 = \sqrt{2}$ and $P_4 P_2 = 2\sqrt{2}$, in Figure 25 we have

$$OL = NP_2 = \tfrac{1}{2} P_2 P_8 = \tfrac{\sqrt{2}}{2}, \quad \text{and} \quad ON = LP_2 = \tfrac{1}{2} P_2 P_4 = \sqrt{2}.$$

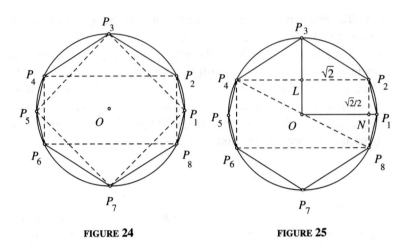

FIGURE 24   **FIGURE 25**

Hence

$$\text{the altitude } LP_3 \text{ of } \triangle P_2 P_3 P_4 = r - OL = \tfrac{\sqrt{5}}{\sqrt{2}} - \tfrac{\sqrt{2}}{2},$$

making

$$\triangle P_2 P_3 P_4 = \tfrac{1}{2} \cdot 2\sqrt{2} \cdot \left(\tfrac{\sqrt{5}}{\sqrt{2}} - \tfrac{\sqrt{2}}{2}\right) = \sqrt{5} - 1.$$

Similarly,

$$\text{the altitude } NP_1 \text{ of } \triangle P_2 P_8 P_1 = r - ON = \tfrac{\sqrt{5}}{\sqrt{2}} - \sqrt{2},$$

and

$$\triangle P_2 P_8 P_1 = \tfrac{1}{2} \cdot \sqrt{2} \cdot \left(\tfrac{\sqrt{5}}{\sqrt{2}} - \sqrt{2}\right) = \tfrac{\sqrt{5}}{2} - 1.$$

Finally, then, observing that $\triangle P_2 P_4 P_8 = \tfrac{1}{2}$ the rectangle $= 2$, the maximum area of the octagon

$$= 2\left[\sqrt{5} - 1 + 2 + \tfrac{\sqrt{5}}{2} - 1\right] = 2\left(\tfrac{3}{2}\sqrt{5}\right) = 3\sqrt{5}.$$

15. (Two Gems from the 2001 Putnam Competition (February, 2002, pages 72–78))

We follow the published solutions. First an intriguing Diophantine equation.

1. Prove that there are unique positive integers a and n such that

$$a^{n+1} - (a + 1)^n = 2001.$$

Useful information is often squeezed out of an equation of this sort by considering it modulo a suitable integer. Since the sum of the digits of 2001 is 3, then 2001 is divisible by 3, and we have (mod 3) that

$$a^{n+1} - (a+1)^n \equiv 0.$$

Clearly this fails for $a \equiv 0$ or -1, implying $a \equiv 1$ and then $a+1 \equiv -1$. Hence

$$a^{n+1} - (a+1)^n \equiv 1 - (-1)^n \equiv 0,$$

from which it follows that n must be even. Although this turns out to be important, on its own it is not nearly enough to lead directly to a solution. The key is not to stop at modulo 3 but to also consider the equation modulo a and modulo $a + 1$.

Working $(\bmod\, a)$, we find that

$$2001 \equiv a^{n+1} - (a+1)^n \equiv 0 - 1,$$

giving $2002 \equiv 0$, which reveals that a must be a divisor of 2002.

Similarly, in $(\bmod\, a + 1)$, we have

$$2001 \equiv a^{n+1} - (a+1)^n \equiv (-1)^{n+1} - 0 \equiv -1$$

(recall n is even), showing that $a + 1$ is also a divisor of 2002.

Now, the prime decomposition of 2002 is $2 \cdot 7 \cdot 11 \cdot 13$. Thus two consecutive divisors could only be 1 and 2 or 13 and 14. Since $a^{n+1} - (a+1)^n$ is negative for $a = 1$, it follows that, if the equation has a solution, a must be 13. Thus, letting $n = 2k$, it remains to solve

$$13^{2k+1} - 14^{2k} = 2001,$$

that is,

$$13(13^2)^k - (14^2)^k = 2001.$$

Recalling that, modulo 8, the square of an odd number is congruent to 1 (odd numbers are $8k + r$, where $r = 1, 3, 5,$ or 7, and $1^2, 3^2, 5^2, 7^2$, are all $\equiv 1 \pmod 8$), there is a chance that it might be to our advantage to consider this equation (mod 8). Doing so, we find that

$$13 \cdot 1 - 196^k \equiv 1,$$
$$5 - 4^k \equiv 1,$$

and

$$4^k \equiv 4.$$

Now, if $k > 1$, then 4^k is divisible by 8, that is, $4^k \equiv 0$, not 4, and we conclude that, if a solution exists, k must be 1 and $n = 2$. Thus it remains only to check whether

$$13^3 - 14^2 = 2001.$$

As expected, our hopes are confirmed by direct calculation:

$$13^3 - 14^2 = 2197 - 196 = 2001.$$

Now for a lovely combinatorics problem.

2. Let n be an even positive integer. Write the numbers $1, 2, \ldots, n^2$ in order in the squares of an $n \times n$ grid G, going row by row from the top and from left to right in each row. Thus the first row is $(1, 2, \ldots, n)$, the second row is $(n + 1, n + 2, \ldots, 2n)$, and so on. Now color each square either red or black so that exactly half the squares in each row are red and also exactly half the squares in each column are red (a checkerboard coloring is one possibility).

   For all such colorings of G, prove that the sum of the numbers in all the red squares is the same as the sum of the numbers in all the black squares.

   It is evident that the second row of G is obtained by adding n to each member of the first row, that the third row is obtained by adding $2n$ to each member of the first row, and so on through the rows. Although a simple consequence of this property, it is very perceptive to realize that this permits the following decomposition of G into a sum of two $n \times n$ squares G_1 and G_2:

   let each row of G_1 be the first row of G, namely $1, 2, \ldots, n$, and let each entry in the kth row of G_2 be $(k - 1)n$; then, adding the numbers in corresponding squares, G is obtained as the sum $G_1 + G_2$.

$$G_1 : \begin{bmatrix} 1 & 2 & 3 & \cdots & n \\ 1 & 2 & 3 & \cdots & n \\ 1 & 2 & 3 & \cdots & n \\ \cdots & \cdots & \cdots & \cdots & \\ 1 & 2 & 3 & \cdots & n \end{bmatrix},$$

$$G_2 : \begin{bmatrix} 0 & 0 & 0 & \cdots & 0 \\ n & n & n & \cdots & n \\ 2n & 2n & 2n & \cdots & 2n \\ \cdots & \cdots & \cdots & \cdots & \\ (n-1)n & (n-1)n & (n-1)n & \cdots & (n-1)n \end{bmatrix}$$

Now let the squares of G_1 and G_2 be colored red or black in keeping with the coloring of G. Then the sum of the numbers in the red squares of G is the sum of the numbers in the red squares of G_1 and the numbers in the red squares of G_2.

Since each row of G_2 contains n copies of the same number and half of these are in red squares and half in black squares, the sum of the numbers in the red squares in a row of G_2 is the same as the sum of the numbers in the black squares in the row. Holding for each row, it follows in G_2 that the sum of the numbers in all the red squares is the same as the sum of the numbers in all the black squares.

Similarly, each column of G_1 contains n copies of the same number, and since half of them are in red squares, it follows from the same line of reasoning that also in G_1 the sum of the numbers in all the red squares is the same as the sum of the numbers in all the black squares. Since this holds in G_1 and G_2, it holds also in their sum G.

16 A Special Set of Rational Points on the Circle $x^2 + y^2 = 1$

(Problem 1637, proposed by Erwin Just (Emeritus) Bronx Community College, Bronx, NY., December, 2002, page 404.)

Prove that the circle with equation $x^2 + y^2 = 1$ contains an infinite number of points with rational coordinates such that the distance between each pair of the points is irrational.

We follow the brilliant solution of Roy Barbara, Lebanese University, Fanar, Lebanon.

Let p be an odd prime of the form $4k + 1$. Then, according to a famous theorem of Fermat, p has a unique unordered expression as the sum of the squares of two positive integers a and $b : p = a^2 + b^2$. Since p is odd, a and b cannot be equal. Let $a > b$. Then the point

$$P\left(\frac{a^2 - b^2}{p}, \frac{2ab}{p}\right)$$

has rational coordinates and is a point on the circle $x^2 + y^2 = 1$:

$$\left(\frac{a^2 - b^2}{p}\right)^2 + \left(\frac{2ab}{p}\right)^2 = \frac{a^4 - 2a^2b^2 + b^4}{p^2} + \frac{4a^2b^2}{p^2}$$

$$= \frac{a^4 + 2a^2b^2 + b^4}{p^2}$$

$$= \frac{p^2}{p^2} = 1.$$

We observe that these coordinates are in their lowest terms: $a^2 - b^2$ is clearly a positive integer that is less than the prime p, and since $a \neq b$, the arithmetic mean–geometric mean inequality gives $2ab < a^2 + b^2 = p$.

Each prime p of the form $4k + 1$, then, gives a rational point P on the circle. Since there is an infinity of such primes, there are an infinite number of rational points like P on the circle.

Let

$$Q\left(\frac{c^2 - d^2}{q}, \frac{2cd}{q}\right)$$

be the point given by a different prime q of the form $4k + 1$. Then Q can't be the same point as P since their irreducible coordinates have different denominators. Recalling that the coordinates of P and Q satisfy the equation $x^2 + y^2 = 1$, the nonzero distance PQ is

$$PQ = \sqrt{\left(\frac{a^2 - b^2}{p} - \frac{c^2 - d^2}{q}\right)^2 + \left(\frac{2ab}{p} - \frac{2cd}{q}\right)^2}$$

$$= \sqrt{2 - \frac{2(a^2 - b^2)(c^2 - d^2)}{pq} - \frac{8abcd}{pq}}$$

$$= \frac{1}{\sqrt{pq}}\sqrt{2pq - 2(a^2 - b^2)(c^2 - d^2) - 8abcd}$$

$$= \frac{1}{\sqrt{pq}}\sqrt{2(a^2 + b^2)(c^2 + d^2) - 2(a^2 - b^2)(c^2 - d^2) - 8abcd}$$

$$= \frac{1}{\sqrt{pq}}\sqrt{4(a^2d^2 - 2abcd + b^2c^2)}$$

$$= \frac{2|ad - bc|}{\sqrt{pq}},$$

which is irrational since the numerator is an integer and pq is not a perfect square.

17. (Two Pretty Problems from the 2002 Putnam Competition (Feb., 2003, page 76, 77.))

    1. Given any five points on a sphere, show that some four of them must lie on a closed hemisphere.

    2. Let $n \geq 2$ be an integer and let T_n be the number of non-empty subsets S of $\{1, 2, 3, \ldots, n\}$ with the property that the average of the elements of S is an integer. Prove $T_n - n$ is always an even number.

Solutions

    1. A great circle C through two of the given points partitions the sphere into two closed hemispheres. Of the other three points, at least two must lie on the same side of C, possibly on C itself, for a total of at least four points in that closed hemisphere.

    2. Besides the n singleton subsets $\{1\}, \{2\}, \{3\}, \ldots, \{n\}$, there are $T_n - n$ other subsets that are counted by T_n, each of which has the property that the average of its members is an integer.

        Let A be the class of these other subsets which contain their average as a member of the set, and let B be the class of subsets which don't. Since the addition or deletion of the average of the members of a subset does not change the average of the resulting subset, there is a one-to-one correspondence between A and B:

        deleting the average from a subset in A yields a subset in B, and adding the average to a subset in B gives a subset in A.

    Thus A and B have the same cardinality, making $T_n - n$ an even number.

Appendix (i)

1. Let P be a point in the plane of a circle C. P might be on, inside, or outside C. Let a straight line through P intersect C at A and B (Figure 26). Then the product of the directed segments $PA \cdot PB$ is called the *power* of P with respect to C. This is well-defined, for it is an easy exercise in similar triangles to show that $PA \cdot PB$ is independent of the line through P. We observe that the power of a point outside C is positive, the power of a point on C is zero, and the power of a point inside C is negative.

2. The locus of a point whose powers with respect to circles C and D are equal is called the *radical axis* of C and D.

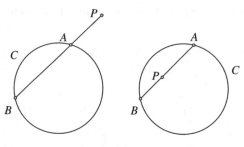

FIGURE 26

The radical axis of two *intersecting* circles is therefore the line of their common chord (Figure 27) (using the fact that the power is negative for a point inside a circle and positive for a point outside, it is another straightforward exercise to show that a point lies on the radical axis if and only if it is on the line of their common chord).

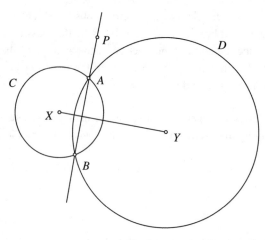

FIGURE 27

Since the centers of intersecting circles are each equidistant from the points of intersection (in Figure 27, $XA = XB$ and $YA = YB$), the segment which joins the centers (XY) is the perpendicular bisector of the common chord (AB). Consequently, the radical axis of two intersecting circles is given by the line through a point of intersection (A or B) which is perpendicular to the line joining the centers (XY). This is precisely the situation in the problem in part 3 of this section.

3. If C, D, E are circles which intersect in pairs and whose centers are not collinear, then no two of the radical axes of the three pairs are parallel.

Let the radical axis of (C, D) meet the radical axis of (D, E) in the point R. Then R has the same power with respect to all three circles C, D, E, and accordingly is a point on the radical axis of the pair (C, E). That is to say, the three radical axes of C, D, and E, taken in pairs, are concurrent, and the point of concurrency is appropriately called the *radical center* of the circles.

(ii)

To show that $AH = 2 \cdot OM$, let BO be extended to meet the circumcircle at G (Figure 28).

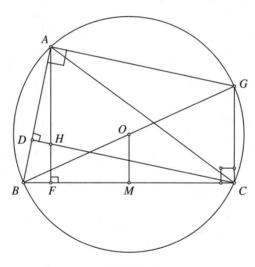

FIGURE 28

Now, M is actually the midpoint of BC, and therefore OM joins the midpoints of two sides of $\triangle BCG$, implying that CG is parallel to OM and twice as long. Since OM is perpendicular to BC, as is altitude AF, it follows that AH, OM, and CG are all parallel. But so are AG and CH parallel since both are perpendicular to AB (CHD is an altitude, and $\angle BAG = 90°$ because BG is a diameter). Thus the pairs of opposite sides of quadrilateral $AHCG$ are parallel, making it a parallelogram. Its opposite sides AH and CG are therefore equal, and since CG is twice OM, so is AH.

From *The College Mathematics Journal*

1 A Novel Way of Adding and Multiplying

In their note "A Picture for Real Arithmetic," Paul Fielstad and Peter Hammer made the following delightful observations on how adding and multiplying can be performed geometrically (Jan. 2000, page 56–60).

Let the real number line, in the form of the x-axis in a coordinate plane, be mapped to the unit circle by stereographic projection from the point $(0,1)$ (Figure 1). The image of the number x is the point on the unit circle at which it is crossed by the line that joins $(x, 0)$ and the "pole" $(0, 1)$. Thus each number x lends its name to a point on the circle. The intercepts 1 and -1 retain their identities, the numbers > 1 map to the arc in the first quadrant, the numbers < -1 to the arc in the second quadrant, and the numbers between 1 and -1 to the arc below the x-axis.

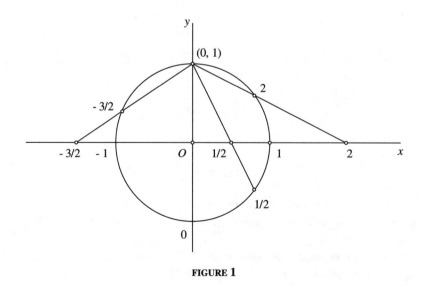

FIGURE 1

Let L be the line which joins the numbers a and b on the circle. Then the product of a and b is found simply by projecting the y-intercept P of L onto the circle from the point 1 (Figure 2); the sum of a and b is similarly found by projecting to the circle, this time from the point 0, the point of intersection Q of L and the tangent $y = 1$ (Figure 3).

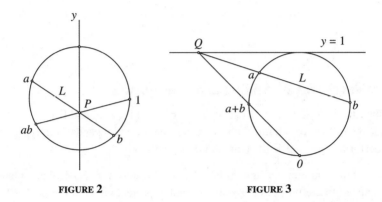

FIGURE 2   FIGURE 3

We observe that this provides a nice geometric demonstration that the product of two negative numbers is positive (Figure 4) and that the product of a negative number and a positive number is negative (Figure 2).

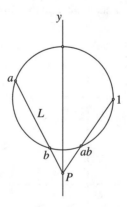

FIGURE 4

The proofs are straightforward derivations in basic analytic geometry. Therefore let us prove only the rule for multiplication and leave the addition rule as an exercise.

We begin by finding the coordinates (X, Y) of the point on the circle that represents the number t on the x-axis (Figure 5).

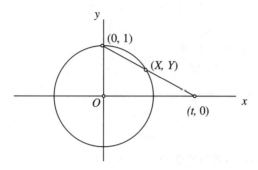

FIGURE 5

The equation of the circle is $x^2 + y^2 = 1$ and the line through $(0, 1)$ and $(t, 0)$ is $x + ty - t = 0$. Solving we obtain

$$(X, Y) = \left(\frac{2t}{t^2 + 1}, \frac{t^2 - 1}{t^2 + 1} \right).$$

Next let's determine the y-intercept of the line which joins the points representing c and d on the circle. These points have coordinates

$$\left(\frac{2c}{c^2 + 1}, \frac{c^2 - 1}{c^2 + 1} \right) \quad \text{and} \quad \left(\frac{2d}{d^2 + 1}, \frac{d^2 - 1}{d^2 + 1} \right).$$

The slope of the line joining them is

$$\frac{\frac{c^2-1}{c^2+1} - \frac{d^2-1}{d^2+1}}{\frac{2c}{c^2+1} - \frac{2d}{d^2+1}} = \frac{(c^2 - 1)(d^2 + 1) - (d^2 - 1)(c^2 + 1)}{2[c(d^2 + 1) - d(c^2 + 1)]}$$

$$= \frac{2c^2 - 2d^2}{2(cd^2 - dc^2 + c - d)} = \frac{c^2 - d^2}{cd(d - c) + (c - d)} = -\frac{c+d}{cd-1}.$$

Thus the equation of the line joining them is

$$y - \frac{c^2 - 1}{c^2 + 1} = -\frac{c + d}{cd - 1}\left(x - \frac{2c}{c^2 + 1} \right),$$

and the y-intercept is

$$y = \frac{c^2 - 1}{c^2 + 1} + \frac{c + d}{cd - 1} \cdot \frac{2c}{c^2 + 1}$$

$$= \frac{1}{c^2 + 1}\left(c^2 - 1 + \frac{2c(c + d)}{cd - 1} \right)$$

$$= \frac{1}{c^2 + 1} \cdot \frac{c^3 d - c^2 - cd + 1 + 2c^2 + 2cd}{cd - 1}$$

$$= \frac{c^3 d + c^2 + cd + 1}{(c^2 + 1)(cd - 1)} = \frac{(c^2 + 1)(cd + 1)}{(c^2 + 1)(cd - 1)} = \frac{cd + 1}{cd - 1}.$$

Thus the y-intercept of the line joining a and b on the circle is

$$\frac{ab + 1}{ab - 1},$$

and the y-intercept of the line joining ab to 1 is

$$\frac{(ab) \cdot 1 + 1}{(ab) \cdot 1 - 1} = \frac{ab + 1}{ab - 1}.$$

That is to say, the segments joining a to b and ab to 1 intersect on the y-axis, and therefore the projection from the number 1 in the rule for multiplication is established.

2 How to Construct a Tangent to an Ellipse from a Point Outside

(a) In the Miscellania on page 317 of the September issue, 2001, David Bloom of Brooklyn College gives the following method of drawing a tangent to an ellipse E from a point P outside (Figure 6):

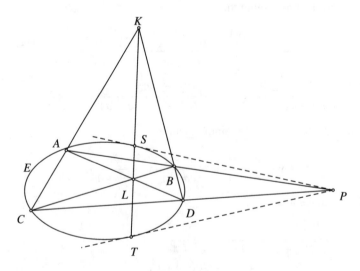

FIGURE 6

Through P draw any two secants PBA and PDC such that CA and DB are not parallel. Let CA meet DB at K and let CB meet AD at L. Then KL meets E at the points of contact S and T of the tangents from P.

Professor Bloom writes that he found this so beautiful when he learned of it in 1958 that it has warmed his heart ever since.

The proof is connected with poles and polars in projective geometry and an essay like this is not the place for an organized treatment of this subject. However, it is such a marvelous discovery that I couldn't resist sharing at least the construction with you.

(b) Another approach is given in the delightful Classroom Capsule "A Quick Construction of Tangents to an Ellipse" by Arthur C. Segal of the University of Alabama at Birmingham (March, 2000, page 131), where it is shown that drawing two arcs and two tangents to a circle is all you need to do!

Let us draw a tangent to the arc of the ellipse

$$\frac{x^2}{a^2} + \frac{y^2}{b^2} = 1$$

which lies in the first quadrant. We assume its axes are known.

First draw the first-quadrant circular arcs $O(a)$ and $O(b)$, that is, with center the origin and radii equal to the semi-axes a and b. Then, from any point A on $O(a)$, draw AO to give B on $O(b)$. Let the tangents to the circles at A and B meet the x and y axes at T and S as in Figure 7. Then ST is a tangent to the ellipse.

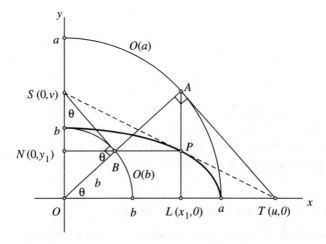

FIGURE 7

Let AL be the ordinate of A and suppose $OL = x_1$ (Figure 7). Let the line through B which is parallel to the x-axis meet AL at P and the y-axis at $N(0, y_1)$. Thus AP and BP are perpendicular and $PL = y_1$; in fact, this makes P the point (x_1, y_1). Let S and T be the points $(0, v)$ and $(u, 0)$. Finally, let $\angle AOL = \theta$. We note that, at this point, we do not know that P lies on the ellipse.

Then $\angle AOS = 90° - \theta$, implying $\angle OSB = \theta$ in right triangle OSB; also, NB is parallel to the x-axis and alternate angle $NBO = \theta$. Now, from right triangles AOL and AOT, we have that

$$\cos \theta = \frac{x_1}{a} = \frac{a}{u},$$

and from right triangles NOB and OSB we have

$$\sin \theta = \frac{y_1}{b} = \frac{b}{v}.$$

Therefore

$$\frac{x_1{}^2}{a^2} + \frac{y_1{}^2}{b^2} = \cos^2 \theta + \sin^2 \theta = 1,$$

and we conclude that the point $P(x_1, y_1)$ does lie on the ellipse.

From the values of $\cos \theta$ and $\sin \theta$ we also obtain

$$u = \frac{a^2}{x_1} \quad \text{and} \quad v = \frac{b^2}{y_1}.$$

Now, the intercept form of the equation of ST is

$$\frac{x}{u} + \frac{y}{v} = 1,$$

that is,

$$\frac{x}{\frac{a^2}{x_1}} + \frac{y}{\frac{b^2}{y_1}} = \frac{x_1 x}{a^2} + \frac{y_1 y}{b^2} = 1,$$

which we recognize as the tangent to the ellipse at the point $P(x_1, y_1)$.

We observe that P is the point where the ordinate from A crosses the ellipse. Thus P may be specified in advance and A determined from P by simply extending the ordinate of P to $O(a)$.

3 A Note on Euler's Congruent Numbers

This note is based on the outstanding article "Three Fermat Trails to Elliptic Curves" by Bud Brown of Virginia Tech (May 2000, pages 162–172).

Euler defined a "congruent number" to be a rational number that is the area of a right-angled triangle which has rational sides; that is to say, if (x, y, z) is a triple of rational numbers such that $x^2 + y^2 = z^2$, then $xy/2$ is a congruent number.

All it takes to make $xy/2$ rational is x and y to be rational; the sticky point is arranging $x^2 + y^2$ to be the square of a rational number. However, since integers are rational, congruent numbers are as common as Pythagorean triples. For example, the $(3, 4, 5)$ triangle yields the congruent number 6. It turns out that not every rational number is a congruent number, and therefore it is a matter of interest just which ones are and which ones aren't.

Fermat showed that the difference between the fourth powers of two integers is never a perfect square (a nice application of his method of infinite descent) and it is easy to see that if it's true for integers, it's also true for rational numbers (multiplying by the fourth power of a common denominator of such rational numbers would convert them to integers). This eliminates from the set of congruent numbers all the squares of rational numbers:

if rational numbers x, y, z, satisfy $x^2 + y^2 = z^2$, and their congruent number were to be a rational square, $xy/2 = w^2$, then $4x^2y^2 = 16w^4$ and $(x^2 - y^2)^2 = (x^2 + y^2)^2 - 4x^2y^2 = z^4 - (2w)^4$, contradicting Fermat's theorem.

Thus there is no rational right triangle with area 1, or 4, or $\frac{1}{4}$ ....

A Characterization of Congruent Numbers

A positive rational number n is a congruent number if and only if there exists a rational number u such that $u^2 - n$ and $u^2 + n$ are the squares of rational numbers.

Necessity: Suppose n is a congruent number. Then, for some rational numbers x, y, z, we have $x^2 + y^2 = z^2$ and $xy/2 = n$. In this case,

$$\frac{x+y}{2}, \frac{x-y}{2}, \quad \text{and} \quad n$$

are rational, and

$$\left(\frac{x+y}{2}\right)^2 = \frac{x^2 + 2xy + y^2}{4}$$

$$= \frac{x^2 + y^2}{4} + \frac{xy}{2} = \left(\frac{z}{2}\right)^2 + n;$$

similarly,

$$\left(\frac{x-y}{2}\right)^2 = \left(\frac{z}{2}\right)^2 - n.$$

Setting the rational number $z/2 = u$, then $u^2 - n$ and $u^2 + n$ are the squares of rational numbers.

Sufficiency: Suppose n and u are rational numbers such that $\sqrt{u^2 - n}$ and $\sqrt{u^2 + n}$ are rational. Then

$$x = \sqrt{u^2 + n} + \sqrt{u^2 - n},$$
$$y = \sqrt{u^2 + n} - \sqrt{u^2 - n},$$

and $2u$ are rational numbers such that

$$x^2 + y^2 = 2(u^2 + n) + 2(u^2 - n) = (2u)^2, \quad \text{a rational square.}$$

Also

$$\frac{xy}{2} = \frac{1}{2}[(u^2 + n) - (u^2 - n)] = n, \quad \text{a rational number,}$$

and it follows that n is a congruent number.

This doesn't make it particularly easy to find congruent numbers that are off the beaten track. In hindsight we can see that, for $(x, y, z) = (3, 4, 5)$, we have $n = 6$, and from $z/2 = u$, we get $u = \frac{5}{2}$. Thus $\left(\frac{5}{2}\right)^2 - 6$ and $\left(\frac{5}{2}\right)^2 + 6$ should be rational squares: in fact,

$$\left(\tfrac{5}{2}\right)^2 - 6 = \tfrac{1}{4} \quad \text{and} \quad \left(\tfrac{5}{2}\right)^2 + 6 = \tfrac{49}{4}.$$

In certain cases, however, a Pythagorean triple can lead to a second congruent number. The triple $(9, 40, 41)$ generates the congruent number $\frac{9 \cdot 40}{2} = 180$, which has the square factor 36. Thus, dividing by its square root, 6, we obtain a rational right triangle with sides $(x, y, z) = (\frac{9}{6}, \frac{40}{6}, \frac{41}{6})$, which gives the congruent numbers

$$\frac{1}{2} \cdot \frac{9}{6} \cdot \frac{40}{6} = \frac{180}{36} = 5.$$

Now, in this particular case,

$$u = \frac{z}{2} = \frac{41}{12},$$

and we obtain the three rational squares

$$(u^2 - 5, u^2, u^2 + 5) = \left(\frac{1681}{144} - \frac{720}{144}, \frac{1681}{144}, \frac{1681}{144} + \frac{720}{144}\right)$$

$$= \left(\frac{961}{144}, \frac{1681}{144}, \frac{2401}{144}\right)$$

$$= \left(\left(\frac{31}{12}\right)^2, \left(\frac{41}{12}\right)^2, \left(\frac{49}{12}\right)^2\right),$$

which we observe are in *arithmetic progression.*

For the remarkable relationship between three rational squares in arithmetic progression and elliptic curves, make sure you don't miss reading the gem by Bud Brown referred to at the beginning of this note.

4 Eisenstein Triples

(Based on the article Meta-Problems in Mathematics, by Al Cuoco of the Education Development Center; Nov. 2000, pages 373–378)

Three positive integers (a, b, c) such that $a^2 + b^2 = c^2$ is called a Pythagorean triple. It is easy to see that if any two of a, b, c have a common divisor $d > 1$, then d must also divide the third member: For example, suppose an integer $d > 1$ divides a and b. Then, for some integers m and n, $a = dm$ and $b = dn$, giving

$$c^2 = a^2 + b^2 = d^2 m^2 + d^2 n^2 = d^2 (m^2 + n^2).$$

Hence d^2 divides c^2 and it follows that d divides c.

A triple in which a, b, and c are pairwise relatively prime is called a *primitive* triple. Thus there are only two kinds of Pythagorean triples: (i) primitive triples and (ii) multiples of primitive triples.

Now,

$$(a, b, c) = (x^2 - y^2, 2xy, x^2 + y^2)$$

is a primitive Pythagorean triple if and only if x and y are relatively prime positive integers where $x > y$ and one of them is even and the other odd (this is proved in essay 13 of my Ingenuity in Mathematics, The Anneli Lax New Mathematical Library, vol. 23, 1970). At least it is easy to see that this formula does generate a Pythagorean triple, for clearly

$$(x^2 - y^2)^2 + (2xy)^2 = (x^2 + y^2)^2.$$

Our present interest in this result is simply to observe how one might happen upon it among the most elementary properties of complex numbers.

If $z = x + yi$ is a complex number, recall that $\bar{z} = x - yi$ is the conjugate of z, and the norm of z, denoted by $N(z)$, is given by

$$N(z) = z\bar{z} = (x + yi)(x - yi) = x^2 + y^2.$$

It is a straightforward exercise to show that the norm of a product is the product of the norms:

$$N(z_1 z_2) = N(z_1)N(z_2).$$

It follows, then, that $N(z^2) = N(z)^2$.

If $z = x + yi$, where x and y are positive integers with $x > y$, then

$$z^2 = x^2 - y^2 + 2xyi,$$

and $N(z^2) = N(z)^2$ gives the desired result already:

$$(x^2 - y^2)^2 + (2xy)^2 = (x^2 + y^2)^2.$$

Geometrically, if $a^2 + b^2 = c^2$, there is a triangle having sides of lengths a, b, and c in which the angle opposite side c is a right angle. Ever reminded by gravity of the importance of the directions horizontal and vertical, the properties of right triangles are understandably very close to our hearts. Applying the law of cosines to our right triangle we obtain the confirmation that

$$c^2 = a^2 + b^2 - 2ab\cos 90° = a^2 + b^2.$$

Now, 60° angles are very nice angles, too, and they occur in a multitude of interesting configurations. If a triangle has sides of lengths a, b, and c and a 60° angle opposite side c, then the law of cosines gives

$$c^2 = a^2 + b^2 - 2ab\cos 60° = a^2 + b^2 - ab.$$

A triple of positive integers (a, b, c) such that

$$a^2 + b^2 - ab = c^2$$

is called an "Eisenstein" triple in honor of a student of Gauss. The question is "Is there a formula for Eisenstein triples like the one for Pythagorean triples?"

Remarkably, the problem yields to the simplest properties of the cube roots of unity. Recall that the cube roots of unity are 1, ω, and ω^2, where

$$\omega = \frac{-1 + i\sqrt{3}}{2},$$

and that their sum $1 + \omega + \omega^2 = 0$.

If $z = a + b\omega$, the norm of z is

$$N(z) = (a + b\omega)\overline{(a + b\omega)}.$$

Now, it is again a straightforward exercise to show that the conjugate of a sum is the sum of the conjugates and that the conjugate of a product is the product of the conjugates. Since the conjugate of a real number b is just b itself $(\overline{b + 0i} = b - 0i = b)$, and the conjugate of ω is its conjugate root ω^2, we have

$$
\begin{aligned}
N(z) &= (a + b\omega)\overline{(a + b\omega)} \\
&= (a + b\omega)(a + b\overline{\omega}) \\
&= (a + b\omega)(a + b\omega^2) \\
&= a^2 + b^2\omega^3 + (\omega + \omega^2)\,ab \\
&= a^2 + b^2 - ab \qquad \text{(since } \omega^3 = 1 \text{ and } \omega + \omega^2 = -1).
\end{aligned}
$$

That is to say, $N(a + b\omega) = a^2 + b^2 - ab$.

Thus, if x and y are integers,

$$N(x + y\omega) = x^2 + y^2 - xy = \text{some integer } c,$$

and

$$N[(x + y\omega)^2] = c^2, \text{ the square of an integer.}$$

Consequently, if we can find integers a and b such that

$$(x + y\omega)^2 = a + b\omega,$$

they would give a desired triple (a, b, c) when combined with $c = x^2 + y^2 - xy$:

$$c^2 = N[(x + y\omega)^2] = N(a + b\omega) = a^2 + b^2 - ab.$$

Now,

$$
\begin{aligned}
(x + y\omega)^2 &= x^2 + 2xy\omega + y^2\omega^2 \\
&= x^2 + 2xy\omega + y^2(-1 - \omega) \\
&= (x^2 - y^2) + (2xy - y^2)\omega,
\end{aligned}
$$

and we have

$$a = x^2 - y^2, \quad b = 2xy - y^2.$$

Thus an Eisenstein triple

$$(a, b, c) = (x^2 - y^2, 2xy - y^2, x^2 + y^2 - xy)$$

is given by each pair of integers x and y.

    We note that no claim is made that this generates all Eisenstein triples. While the relation $a^2 + b^2 - ab = c^2$ holds for all integers x and y, if we don't want to lose sight of the sides of $60°$ triangles we must choose $x > y > 0$ in order to keep the integers positive. For example, $x = 3$, $y = 1$ gives the triple $(8, 5, 7)$, for which

$$8^2 + 5^2 - 8 \cdot 5 = 64 + 25 - 40 = 49 = 7^2.$$

5 Two "Mathematics Without Words"

In coming to an appreciation of one of those clever figures known as "mathematics without words" or "proofs without words," sometimes it takes me a while to work out how the diagram was put together before I can see how the argument goes. You needn't worry about having to speculate about the two delightful examples in this section for they are "mathematics without words" *with words*.

    (a) The first is due to Professor Emeritus P. D. Barry of the National University of Ireland (Jan. 2001, page 69). It provides an ingenious proof of the identity

$$\arctan\left(x + \sqrt{1 + x^2}\right) = \frac{\pi}{4} + \frac{1}{2}\arctan x.$$

    Taking an arbitrary length x, construct right triangle ABC with legs 1 and x, thus making the hypotenuse $AC = \sqrt{1 + x^2}$ (Figure 8). Next, let side x be extended to meet the circle with center C and radius AC at D, to make $CD = \sqrt{1 + x^2}$, and $BD = x + \sqrt{1 + x^2}$. Then clearly

$$\angle DAB = \arctan\left(x + \sqrt{1 + x^2}\right).$$

Let $\angle CAB = v$ and $\angle CAD = u$. Then, in isosceles triangle ACD, the base angle at D also equals u and the exterior $\angle ACB = 2u$. In $\triangle ABC$, then, v and $2u$ are complementary, giving

$$2u + v = \frac{\pi}{2}$$

and

$$u = \frac{\pi}{4} - \frac{1}{2}v.$$

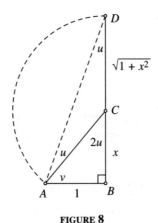

FIGURE 8

Therefore

$$\arctan\left(x + \sqrt{1 + x^2}\right) = \angle DAB = u + v = \frac{\pi}{4} + \frac{1}{2}v,$$

and since $v = \arctan x$, the conclusion follows.

(b) The second comes from Professor Rick Mabry (Jan. 2001, page 19) and is a most charming way of illustrating

$$\frac{1}{3} + \frac{1}{3^2} + \frac{1}{3^3} + \ldots = \frac{1}{2}.$$

Beginning with any rectangle of area 1, slice off the outer thirds, coloring one black and the other white (Figure 9). Then, turning through a right angle,

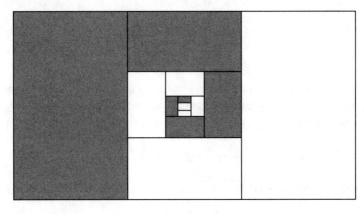

FIGURE 9

do the same thing with the middle third, and continue to do likewise with every resulting middle third.

Clearly the total area colored black is

$$\frac{1}{3} + \frac{1}{3} \cdot \frac{1}{3} + \frac{1}{3} \cdot \frac{1}{3} \cdot \frac{1}{3} + \cdots.$$

But there is an equal white rectangle for every black one, and between them they add up to the whole thing. Voila!

6 A Curious Property of a Parabola

(Based on the article "On The Tangent Lines of a Parabola," by Mikko Stenlund, University of Helsinki, Finland, May 2000, pages 194–196.)

Since all parabolas have the same shape, we may deal with the parabola $y^2 = x$ without loss of generality.

Let $P_1(k^2, k)$ and $P_2(t^2, t)$ be two points on the parabola. What is the locus of the point of intersection P of the tangents at P_1 and P_2 as the points slide around the parabola as a pair that are linked together by the condition $k - t = c$, that is, with constant vertical separation?

Differentiating $y^2 = x$, we get $2yy' = 1$, implying that the slope of the tangent at $P(x, y)$ is $1/2y$. Thus the equation of the tangent at P_1 is

$$y - k = \frac{1}{2k}(x - k^2)$$

or

$$x - 2ky + k^2 = 0,$$

and similarly, the equation of the tangent at P_2 is

$$x - 2ty + t^2 = 0.$$

Solving, we find that P is the point $(kt, (k + t)/2)$. We note in passing the engaging result that P is halfway between the vertical levels of P_1 and P_2.

Now, since $k - t = c$, then $k = t + c$ and the coordinates of P are

$$x = kt = t^2 + ct \quad \text{and} \quad y = \frac{k + t}{2} = t + \frac{c}{2}.$$

Hence, eliminating t, we have

$$t = y - \frac{c}{2},$$

and

$$x = \left(y - \frac{c}{2}\right)^2 + cy - \frac{c^2}{2},$$
$$= y^2 - cy + \frac{c^2}{4} + cy - \frac{c^2}{2},$$
$$= y^2 - \frac{c^2}{4},$$

giving

$$y^2 = x + \frac{c^2}{4}.$$

Thus we have the surprising result that the locus of P is just a copy of the parabola itself that is translated in the direction of the negative x-axis a distance $c^2/4$.

7 A Neat Proof of Heron's Formula

(a) (Based on "Heron's Formula via Proofs Without Words," by Roger B. Nelsen of Lewis and Clark University, Portland, Oregon, Sept. 2001, pages 290–291.)

In the first century A.D., Heron of Alexandria discovered his remarkable formula for the area of a triangle in terms of the lengths of the sides a, b, c:

$$\Delta = \sqrt{s(s-a)(s-b)s-c)}, \text{ where } s = \frac{a+b+c}{2},$$

the semiperimeter of the triangle.

It is well known that the bisectors of the angles of a triangle meet at its incenter and that the area of a triangle is given by the product of its inradius r and semiperimeter s:

$$\Delta = rs.$$

Since the two tangents to a circle from a point outside have the same length, the points of contact of the incircle divide the sides a, b, c, into segments of lengths x, y, z as shown in Figure 10, where angles A, B, and C are denoted by 2α, 2β, and 2γ.

Thus

$$x + y = a, y + z = b, z + x = c, \quad \text{and} \quad s = x + y + z.$$

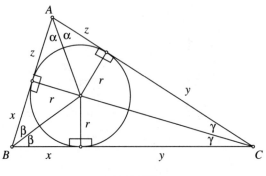

FIGURE 10

Therefore

$$s - a = z, s - b = x, \quad \text{and} \quad s - c = y.$$

In Figure 10, we observe that

$$2\alpha + 2\beta + 2\gamma = 180°, \quad \text{and} \quad \alpha + \beta + \gamma = 90°.$$

Now for Nelsen's coup de grâce, a beautiful way of showing, for angles α, β, γ that add up to a right angle, that

$$\tan \alpha \tan \beta + \tan \beta \tan \gamma + \tan \gamma \tan \alpha = 1.$$

Consider the rectangle $HKLM$ with sides 1 and $\tan \alpha + \tan \beta$ (Figure 11). At K let $\angle HKN = \alpha$; the other angles at K are not known to be β and γ at this

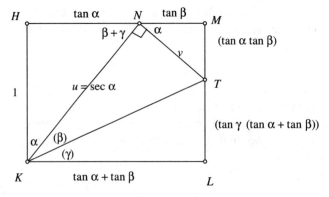

FIGURE 11

point and therefore are marked in parentheses. In $\triangle HKN$, then, we have three things:

$$HN = \tan\alpha, \text{ and hence } NM = \tan\beta;$$

$$u = KN = \sec\alpha,$$

and the angle at N is the complement $\beta + \gamma$ of α.

Next, construct right angle KNT, thus making $\angle MNT = \alpha$. Then, in $\triangle NMT$,

$$\sec\alpha = \frac{NT}{NM} = \frac{v}{\tan\beta}, \quad \text{giving } v = \sec\alpha\tan\beta.$$

Then, in right triangle NKT,

$$\tan\angle NKT = \frac{v}{u} = \frac{\sec\alpha\tan\beta}{\sec\alpha} = \tan\beta,$$

implying $\angle NKT$ is indeed β, from which we also get $\angle TKL = \gamma$.

We note that we can now stop worrying whether the point T might spoil the figure by occurring outside the rectangle on ML extended; since $\alpha + \beta < 90°$ at vertex K, it is clear that KT lies inside the rectangle. Therefore, in $\triangle NMT$, we have

$$\tan\alpha = \frac{MT}{NM} = \frac{MT}{\tan\beta}, \quad \text{giving } MT = \tan\alpha\tan\beta,$$

and in $\triangle KTL$, we have

$$\tan\gamma = \frac{TL}{KL} = \frac{TL}{\tan\alpha + \tan\beta}, \quad \text{giving } TL = \tan\gamma(\tan\alpha + \tan\beta).$$

Adding MT and TL, then, gives the desired

$$\tan\alpha\tan\beta + \tan\beta\tan\gamma + \tan\gamma\tan\alpha = 1.$$

Going back to Figure 10, we find

$$\tan\alpha = \frac{r}{z}, \tan\beta = \frac{r}{x}, \tan\gamma = \frac{r}{y}.$$

Thus

$$1 = \tan\alpha\tan\beta + \tan\beta\tan\gamma + \tan\gamma\tan\alpha$$

$$= \frac{r}{z}\cdot\frac{r}{x} + \frac{r}{x}\cdot\frac{r}{y} + \frac{r}{y}\cdot\frac{r}{z} = r^2\left(\frac{1}{xz} + \frac{1}{xy} + \frac{1}{yz}\right)$$

$$= \frac{r^2(x + y + z)}{xyz} = \frac{r^2 s}{xyz} = \frac{r^2 s^2}{sxyz}$$

$$= \frac{(\triangle ABC)^2}{s(s - a)(s - b)(s - c)},$$

giving $(\Delta ABC)^2 = s(s-a)(s-b)(s-c)$, and Heron's marvelous formula

$$\Delta = \sqrt{s(s-a)(s-b)(s-c)}.$$

(b) Thanks to James Tanton of The Math Circle, I am now able to share with you an approach to Heron's formula that is surely the ultimate in simplicity!

From the well known formula $\Delta = \frac{1}{2}ab \sin C$, we obtain

$$\sin C = \frac{2\Delta}{ab};$$

and from the law of cosines, we have

$$\cos C = \frac{a^2 + b^2 - c^2}{2ab}.$$

Substituting in $\sin^2 C + \cos^2 C = 1$, we get

$$1 = \frac{4\Delta^2}{a^2 b^2} + \frac{(a^2 + b^2 - c^2)^2}{4a^2 b^2},$$

which, when simplified, is Heron's formula already.

Clearing of fractions, we obtain

$$4a^2 b^2 = 16\Delta^2 + a^4 + b^4 + c^4 + 2a^2 b^2 - 2a^2 c^2 - 2b^2 c^2,$$
$$16\Delta^2 = 2(a^2 b^2 + a^2 c^2 + b^2 c^2) - (a^4 + b^4 + c^4).$$

Thus we would like to show that

$$16s(s-a)(s-b)(s-c) = 2(a^2 b^2 + a^2 c^2 + b^2 c^2) - (a^4 + b^4 + c^4),$$

that is,

$$2s(2s - 2a)(2s - 2b)(2s - 2c) = 2(a^2 b^2 + a^2 c^2 + b^2 c^2) \\ - (a^4 + b^4 + c^4),$$

or

$$(a + b + c)(b + c - a)(a + c - b)(a + b - c) = 2(a^2 b^2 + a^2 c^2 + b^2 c^2) \\ - (a^4 + b^4 + c^4)$$

which is easily verified directly.

From the *Pi Mu Epsilon Journal*

1 A Noteworthy Property of a Triangle

(Based on the paper Bisecting a Triangle, by Anthony Todd, Abilene Christian University, Fall, 1999, 31–37)

If a straight line bisects both the perimeter and the area of a triangle, prove that it must pass through the incenter of the triangle.

Clearly a straight line across a triangle can't help meeting at least one of the angle bisectors (Figure 1(a)). Suppose a perimeter and area bisecting line of $\triangle ABC$ crosses BC at D, AC at E, and meets the bisector of angle A at P (Figure 1(b)). Then P is the same distance, say s, from sides AB and AC and is some distance t from BC. Suppose D and E divide the sides BC and AC into parts u, v, x, y, as in the figure. Then, since DE bisects the area, we obtain, from the triangles on each side of DE, that

$$\tfrac{1}{2}(xs + AB \cdot s + ut) = \tfrac{1}{2}(ys + vt), \tag{1}$$

and since DE bisects the perimeter, that

$$x + AB + u = y + v. \tag{2}$$

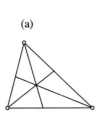
(a)

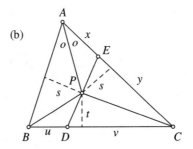

FIGURE 1

From (2), we get $x + AB = y + v - u$. Doubling (1) and substituting for $x + AB$, we obtain

$$(y + v - u)s + ut = ys + vt,$$

and

$$(v - u)s = (v - u)t. \tag{3}$$

Now, if D is *not* the midpoint of BC, then $v - u \neq 0$ and it follows that $s = t$, implying that P is the incenter since it is the same distance from all three sides.

On the other hand, if D *is* the midpoint of BC, then, since a median bisects the area of a triangle, DE would in fact have to be the median to A, implying E and A coincide. Since DE also bisects the perimeter, then sides AB and AC would have to be equal, in which case the triangle would be isosceles and the median DE to the base is easily seen to be the bisector of the vertical angle A, and accordingly passes through the incenter.

To complete the proof, we observe that a virtually identical case is obtained if D and E lie on BC and AC, respectively, and PC is the bisector of angle C (rather than angle A).

Comment   It is not obvious that there exists a line across a triangle that simultaneously bisects its perimeter and its area. Therefore let us conclude this part by giving a proof of this fact.

Let P be a point on a side BC of $\triangle ABC$, and let r be the directed ray from P through the incenter I of the triangle (Figure 2). Let the function $f(P)$ be defined to be (the length of perimeter of $\triangle ABC$ which lies on the left side of r) minus (the length of perimeter on the right side of r) as one stands at P facing I. For some position P, suppose r crosses the perimeter again at Q.

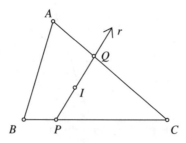

FIGURE 2

It is not difficult to see that f is a continuous function as P moves around the perimeter:

> As P moves to P' (Figure 3), Q moves to Q' and f changes by $(PP' - QQ')$. Clearly, by keeping P' close enough to P, both PP' and QQ' can be made as small as desired, giving a change in f that is smaller in magnitude than any preassigned $\epsilon > 0$.

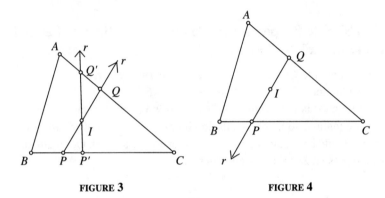

FIGURE 3   FIGURE 4

Now, as P moves around the perimeter of $\triangle ABC$ to Q, the value of $f(Q) = -f(P)$ (Figure 4). By the intermediate value theorem, then, at some point T on the perimeter between P and Q we have $f(T) = 0$, at which point TI bisects the perimeter of the triangle.

In Figure 5, suppose PIQ bisects the perimeter, that is,

$$x + AB + u = y + v.$$

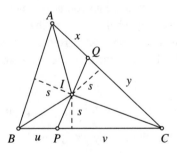

FIGURE 5

Since the incenter I is the same distance (s) from each side, we obtain

$$\tfrac{1}{2}s(x + AB + u) = \tfrac{1}{2}s(y + v),$$

which asserts that PQ also bisects the area of $\triangle ABC$, and the conclusion follows.

It is not difficult to see that this argument, with obvious suitable adjustments, goes through equally well with $f(P)$ defined to be the difference between the *areas* of the parts of the triangle on opposite sides of r, leading to a bisector PIQ of the area of $\triangle ABC$, which is easily shown also to bisect the perimeter of $\triangle ABC$.

2 Two Simple Properties of Would-Be Solutions to Fermat's Equation $x^n + y^n = z^n, n > 2$

(Based on the note *Cranks, Computers, and Fermat's Last Theorem*, by John Zuehlke, Columbia University, Fall, 1999, 45–46)

Although it is now known that there are no positive integer solutions to Fermat's equation, some of us still find it enjoyable to uncover simple conditions that would be required of a solution if one were to exist. Our main interest here is to present Zuehlke's argument that

$$z > \frac{\sqrt[n]{2}}{\sqrt[n]{2} - 1}$$

in any positive integer solution. Admittedly, this is not that impressive a result; for $n = 100$, for example, it only forces z to be greater than 144. But that doesn't matter. It's Zuehlke's pretty proof that is so attractive!

While we're on the subject, let us begin with the easy derivation that each of x and y must exceed n.

(i) $x > n$ and $y > n$: Since the equation is symmetric in x and y, without loss of generality we may label x and y so that $x \geq y$, i.e., $x/y \geq 1$.

  Clearly z must be greater than x. Hence, for some positive integer u we have $z = x + u$. Then, by the binomial theorem,

$$x^n + y^n = z^n = (x + u)^n,$$

$$= x^n + nx^{n-1}u + \text{ a positive quantity},$$

giving $y^n > nx^{n-1}u$. Dividing by y^{n-1}, then

$$y > n \left(\frac{x}{y}\right)^{n-1} u,$$

where each of x/y and u is at least unity. Hence $y > n$, and since $x \geq y$, we also have $x > n$.

(ii) $z > \dfrac{\sqrt[n]{2}}{\sqrt[n]{2} - 1}$:

The equation may be written

$$y = \sqrt[n]{z^n - x^n},$$

which, for a fixed value of z, gives y as a single-valued function of x. Zuehlke now directs attention to the section of the graph of this function that lies in the first quadrant (Figure 6).

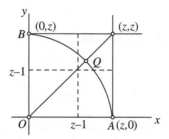

FIGURE 6

Clearly the function is strictly decreasing as x goes from 0 to z, that is, from $B(0, z)$ to $A(z, 0)$. Since Fermat's equation becomes $2x^n = z^n$ for $x = y$, the graph intersects the line $y = x$ at the point

$$Q\left(\frac{z}{\sqrt[n]{2}}, \frac{z}{\sqrt[n]{2}}\right).$$

Since the graph is strictly decreasing from B to Q (the shape of the graph is irrelevant), the ordinate of each point on this part of the curve is greater than the ordinate $z/\sqrt[n]{2}$ of Q. Also, it is evident that the abscissa of each point of the curve between Q and A exceeds the abscissa $z/\sqrt[n]{2}$ of Q.

Observing the dotted lines in Figure 6, it follows that if $z - 1$ were to be less than $z/\sqrt[n]{2}$, either the ordinate or the abscissa of each point of the graph would be between the integers $z - 1$ and z, and thus fail to be an integer itself. Hence, in order for the graph to pass through a lattice point, that is, for Fermat's equation to have a positive integer solution, it must be that

$$z - 1 \geq \frac{z}{\sqrt[n]{2}}.$$

However, since $z/\sqrt[n]{2}$ is irrational, it cannot be equal to the integer $z - 1$, and it must be that

$$z - 1 > \frac{z}{\sqrt[n]{2}},$$

which easily unravels to

$$z > \frac{\sqrt[n]{2}}{\sqrt[n]{2} - 1}.$$

(iii) Since $x > n$ and $y > n$ in any positive integer solution of Fermat's equation, it follows that $z^n = x^n + y^n > n^n + n^n = 2n^n$, giving

$$z > n\sqrt[n]{2}.$$

Zuehlke closes his note by proposing the exercise

(a) Prove

$$\frac{\sqrt[n]{2}}{\sqrt[n]{2} - 1}$$

is a better lower bound on z than $n\sqrt[n]{2}$.

(b) Confirm this by showing that $n\sqrt[n]{2}$ is asymptotic to n while

$$\frac{\sqrt[n]{2}}{\sqrt[n]{2} - 1}$$

is asymptotic to the greater value $n/\ln 2$ (recall that "a and b are asymptotic" means $\lim(a/b) = 1$).

Perhaps you would like to spend a moment in response before reading on.

(a) We would like to show that

$$\frac{\sqrt[n]{2}}{\sqrt[n]{2} - 1} > n\sqrt[n]{2},$$

that is, that

$$\frac{1}{\sqrt[n]{2} - 1} > n,$$

$$n\sqrt[n]{2} - n < 1,$$

$$n\sqrt[n]{2} < n + 1,$$

$$\sqrt[n]{2} < 1 + \frac{1}{n},$$

$$2 < \left(1 + \frac{1}{n}\right)^n,$$

which is clearly so since the binomial theorem gives

$$\left(1 + \frac{1}{n}\right)^n = 1 + n \cdot \frac{1}{n} + \text{a positive quantity.}$$

Since these steps are reversible, the conclusion follows.

(b) Obviously, as n gets arbitrarily large,

$$\lim \frac{n\sqrt[n]{2}}{n} = \lim \sqrt[n]{2} = 1,$$

implying $n\sqrt[n]{2}$ is asymptotic to n.

To complete the exercise, observe that $\sqrt[n]{2} = 2^{1/n} = e^{(\ln 2)/n}$. Hence, for

$$x = \frac{\ln 2}{n}, \quad \text{we have} \quad \sqrt[n]{2} = e^x.$$

Thus

$$\frac{\sqrt[n]{2}}{\sqrt[n]{2} - 1} = \frac{e^x}{e^x - 1},$$

and the ratio

$$r = \frac{\sqrt[n]{2}}{\sqrt[n]{2} - 1} \bigg/ \left(\frac{n}{\ln 2}\right)$$

$$= \frac{e^x}{e^x - 1} \bigg/ \left(\frac{1}{x}\right) = \frac{xe^x}{e^x - 1} = \frac{x}{1 - e^{-x}}.$$

Now, as n increases without bound, x approaches 0, and

$$\frac{x}{1 - e^{-x}}$$

approaches the indeterminate $\frac{0}{0}$. Hence, by L'Hospital's rule, we obtain

$$\lim r = \lim \frac{x}{1 - e^{-x}} = \lim \frac{1}{e^{-x}} = \lim e^x = e^0 = 1,$$

and the conclusion follows.

3 A Problem of Some Subtlety

This is a generalization of Robert Gebhardt's Problem 948 (Fall, 1999, 56). Credit for the solution goes to Amy Kuiper, Alma College, Alma, Michigan, and also to the Grand Valley State University Problem Solving Group, Allendale, Michigan. Robert Gebhardt is an outstanding problemist from Hopatcong, New Jersey.

The outside of an $n \times n \times n$ cube is painted red. Then the cube is chopped into n^3 unit cubes. The unit cubes are put into a box and thoroughly mixed up. Then one unit cube is withdrawn at random from the box and given a good throw across a table. What is the probability that the cube comes to rest with a red face on top?

The key is to observe that the face on top at the end of all this activity could be any of the faces—they're all equally likely. Hence the probability it's red is simply

$$\frac{\text{the number of red faces}}{\text{the total number of faces}} = \frac{6n^2}{6n^3} = \frac{1}{n}.$$

4 On the Smarandache Functions $S(n)$ and $Z(n)$

(a) For n a positive integer, the *Smarandache function* $S(n)$ is the smallest positive integer m such that n divides $m!$. Thus, for example, $S(1) = 1$, $S(2) = 2$, $S(6) = 3$, $S(8) = S(12) = 4$, and except for $n = 1$, $S(n) = m$ simply means that n divides $m!$ but not $(m - 1)!$.

A slight generalization of Problem 956 (Spring, 2000, 106, proposed by Charles Ashbacher, Cedar Rapids, Iowa) asks for a proof that the infinite series

$$T = \frac{1}{S(n)} + \frac{1}{S(n^2)} + \cdots + \frac{1}{S(n^k)} + \cdots$$

diverges for all positive integers n. The solution by Kevin P. Wagner, University of South Florida, Largo, is most incisive.

Observe that $(nk)!$ contains k factors that are multiples of n, namely n, $2n, \ldots, kn$. Hence n^k certainly divides $(nk)!$ if it hasn't already succeeded in dividing any smaller factorial, and it follows that $S(n^k) \leq nk$. Thus

$$\frac{1}{S(n^k)} \geq \frac{1}{nk},$$

and

$$T = \sum_{k=1}^{\infty} \frac{1}{S(n^k)} \geq \sum_{k=1}^{\infty} \frac{1}{nk} = \frac{1}{n} \sum_{k=1}^{\infty} \frac{1}{k},$$

that is, $1/n$ times the harmonic series, which is well known to diverge.

(b) For n a positive integer, the function $Z(n)$, called the *Pseudo-Smarandache function*, is the smallest positive integer m such that n divides the mth triangular number

$$t_m = \frac{m(m + 1)}{2}.$$

It is easy to see that t_m is just m more than t_{m-1}:

$$\frac{m(m+1)}{2} - \frac{(m-1)m}{2} = \frac{2m}{2} = m.$$

Thus, starting with $t_1 = 1$, succeeding triangular numbers are obtained by adding 2, then adding 3, then adding 4, and so forth: 1, 3, 6, 10, 15, 21, 28, 36, 45, 55, 66, ... Hence, for example,

$$Z(2) = 3, \quad Z(3) = 2, \quad Z(4) = 7, \quad Z(5) = 4, \quad Z(15) = 5.$$

Our subject in this section is Problem 961 (Spring, 2000, 109) another proposal by Charles Ashbacher. The solution is due to Paul S. Bruckman, Berkeley, California. The problem is

(i) Prove there are an infinity of integers n such that $Z(n) = n/3$.
(ii) Prove that there are an infinite number of pairs (m, n), $m \neq n$, such that $m \cdot Z(n) = n \cdot Z(m)$.

(i) Observe that $Z(15) = 5$, that $Z(33) = 11$, and that 5 and 11 are two of the infinity of prime numbers $\equiv -1 \pmod 6$. Let's try to show that $Z(3p) = p$ for every prime $p = 6u - 1$.
　　Since

$$\frac{p+1}{2} = 3u,$$

the pth triangular number is divisible by $3p$:

$$t_p = \frac{p(p+1)}{2} = 3pu.$$

Hence $Z(3p) \leq p$, and it remains to show that $3p$ fails to divide each triangular number less than t_p.
　　It is easy to see that $3p$ fails to divide t_{p-1} since

$$t_{p-1} = \frac{(p-1)p}{2} = (3u - 1)(6u - 1)$$

is not even divisible by 3. Moreover, for $k \leq p - 2$, we have

$$2 \cdot t_k = k(k+1),$$

where each factor on the right side is less than p and therefore not divisible by the prime p. Hence $2t_k$ can't be divisible by p, and so neither can t_k. Thus t_k is not divisible by $3p$, and it follows that $Z(3p)$ is indeed equal to p.

(ii) Part (ii) follows trivially from part (i). If $m = 3p, n = 3q$, where p and q are any two unequal members of the infinity of primes $\equiv -1 \pmod 6$, then

$$m \cdot Z(n) = 3p \cdot q = 3q \cdot p = n \cdot Z(m),$$

as desired.

5 On A Certain Equation in Two Variables

This is problem 968 (proposed by Doru Popescu Anastasiu, Liceul Radu Greceanu, Slatina, Romania; Fall, 2000, 157).

Determine all pairs (x, y) of real numbers such that

$$16x^2 + 21y^2 - 12xy - 4x - 6y + 1 = 0.$$

If we denote the left side of the equation by $f(x, y)$, then $z = f(x, y)$ represents a surface in 3-space, and $f(x, y) = 0$ is its intercept with the xy-plane. This problem offers no resistance to an approach by calculus, as is demonstrated by the nice published solution of Megan Foster of Alma College, Alma, Michigan. The following clever solution, due to Soumya kanti Das Bhaumik, Angelo State University, shows that a less sophisticated approach can also succeed.

In dealing with quadratics, it is often useful to complete the square. In the case of two variables, the procedure is to focus on one of the variables and let the chips fall where they may. Since $16x^2$ is already a square, which $21y^2$ isn't, the obvious choice is x, and so let us try to determine a trinomial in x and y whose square yields the three terms in x that occur in the given equation, namely $16x^2$, $-12xy$, and $-4x$. Beginning with $4x$ (whose square would give the $16x^2$), a term of $-\frac{3}{2}y$ would be needed to produce the desired $-12xy$ as a cross-product: $-12xy = 2(4x)(-\frac{3}{2}y)$. Similarly, a term of $-\frac{1}{2}$ would provide the desired cross-product $-4x$ as $2(4x)(-\frac{1}{2})$. Therefore all the terms in x are collected in the square

$$\left(4x - \tfrac{3}{2}y - \tfrac{1}{2}\right)^2 = 16x^2 + \tfrac{9}{4}y^2 + \tfrac{1}{4} - 12xy - 4x + \tfrac{3}{2}y.$$

Thus the given equation may be transformed

$$16x^2 + 21y^2 - 12xy - 4x - 6y + 1 = \left(4x - \tfrac{3}{2}y - \tfrac{1}{2}\right)^2 + \tfrac{75}{4}y^2 - \tfrac{15}{2}y + \tfrac{3}{4}$$

$$= \left(4x - \tfrac{3}{2}y - \tfrac{1}{2}\right)^2 + \tfrac{3}{4}(25y^2 - 10y + 1)$$

$$= \left(4x - \tfrac{3}{2}y - \tfrac{1}{2}\right)^2 + \tfrac{3}{4}(5y - 1)^2 = 0.$$

It follows that each of the squares must vanish, which immediately yields $y = \frac{1}{5}$, and then $x = \frac{1}{5}$. Hence the only real pair is $\left(\frac{1}{5}, \frac{1}{5}\right)$.

Soumya kanti Das Bhaumik observed that any number of problems of this sort can be composed by simplifying an equation

$$p(ax + by - c)^2 + q(dx + ey - f)^2 = 0$$

which is made up of positive numbers p and q and the left sides of the equations of two straight lines that are not parallel.

6 A Problem That Grows on You

Problem 975 (Fall, 2000, 162, proposed by Doru Popescu Anastasiu, Liceul Radu Greceanu, Slatina, Romania) seems pretty contrived and might not appear to promise much excitement. However, once you get into it, it is not without its charm. Here's the problem.

Given a positive integer n, determine positive integers $x_1, x_2, \ldots, x_n$ such that

$$x_1 + 2(x_1 + x_2) + 3(x_1 + x_2 + x_3) + \cdots$$
$$+ n(x_1 + x_2 + \cdots + x_n) = \frac{2n^3 + 3n^2 + 7n}{6}.$$

One can hardly fail to notice the similarity of the right side of the equation and the formula for the sum of the squares of the first n positive integers:

$$1^2 + 2^2 + 3^2 + \cdots + n^2 = \frac{n(n + 1)(2n + 1)}{6} = \frac{2n^3 + 3n^2 + n}{6}.$$

In fact, setting all the x's to their minimum possible value of 1, the left side of the equation becomes

$$1^2 + 2^2 + 3^2 + \cdots + n^2 = \frac{2n^3 + 3n^2 + n}{6},$$

which falls short of the right side by a mere $6n/6 = n$. Thus at least one of the x's needs to be increased.

Observe that each x_k appears in the last bracket on the left side of the equation, and since the coefficient of this bracket is n, a unit increase in any x is enough to increase the left side by the deficit n. Furthermore, since all the x_k, except x_n, appear in more than one bracket, a unit increase in any of $x_1, x_2, \ldots, x_{n-1}$ would bring an increase to the left side of more than n; for example, a unit increase in x_{n-1} would increase the left side by $n + (n - 1)$ from the last two brackets.

Since x_n is the only x that appears only in the last bracket, the only way to hold the increase down to the desired n is to increase x_n to 2. Hence the unique solution is $x_k = 1$ for $k \leq n - 1$ and $x_n = 2$.

7 A Specification of the Incenter

(Problem 972, Fall 2000, proposed by Paul S. Bruckman; solution due to Rex H. Wu, Brooklyn, New York.)

Given three non-collinear points, A, B, C, in the complex plane, determine I, the incenter of triangle ABC, as a "weighted average" of these points.

Let the "position" vector from the origin O to a point P in the plane be denoted by **P**. Also, let the vector from point S to point T be denoted by **ST**. Thus, in terms of the position vectors **S** and **T**, we have $\mathbf{ST} = \mathbf{T} - \mathbf{S}$.

Our problem is to determine numerical coefficients k, m, n, such that the vector sum

$$k\mathbf{A} + m\mathbf{B} + n\mathbf{C} = \mathbf{I}.$$

Now, there are several ways we might specify **I** as a sum of vectors that begins at O. We know that IA and IC bisects angles A and C in the triangle, and if CI meets AB at D (Figure 7), then **I** may be expressed as the sum

$$\mathbf{I} = \mathbf{D} + \mathbf{DI}.$$

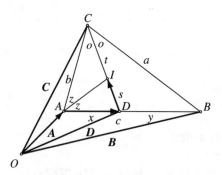

FIGURE 7

As usual, let a, b, c denote the length of the side of $\triangle ABC$ which is opposite the vertex of the same name. Since it is known that a bisector of an angle of a triangle divides the opposite side in the same ratio as the sides about the angle, let the lengths of AD, DB, DI, IC be x, y, s, t, respectively, as in the figure. Then we have

$$\frac{x}{y} = \frac{b}{a}$$

in triangle *ABC* and

$$\frac{s}{t} = \frac{x}{b}$$

in triangle *ADC*. These easily yield

$$\frac{x}{x+y} = \frac{b}{a+b} \quad \text{and} \quad \frac{s}{s+t} = \frac{x}{x+b},$$

that is,

$$\frac{x}{c} = \frac{b}{a+b} \quad \text{and} \quad \frac{s}{DC} = \frac{x}{x+b}.$$

We note that

$$x = \frac{bc}{a+b}.$$

Now,

$$\mathbf{AD} = \frac{x}{c}\mathbf{AB}, \quad \text{where} \quad \mathbf{AB} = \mathbf{B} - \mathbf{A}.$$

Hence

$$\mathbf{AD} = \frac{b}{a+b}(\mathbf{B} - \mathbf{A}).$$

Similarly,

$$\mathbf{DI} = \frac{s}{DC} \cdot \mathbf{DC} = \frac{x}{x+b}(\mathbf{C} - \mathbf{D}).$$

Thus

$$\mathbf{D} = \mathbf{A} + \mathbf{AD} = \mathbf{A} + \left(\frac{b}{a+b}\right)(\mathbf{B} - \mathbf{A}).$$

Now,

$$\begin{aligned}
\mathbf{I} &= \mathbf{D} + \mathbf{DI} \\
&= \mathbf{D} + \frac{x}{x+b}(\mathbf{C} - \mathbf{D}) \\
&= \mathbf{D}\left(1 - \frac{x}{x+b}\right) + \frac{x}{x+b}\mathbf{C} \\
&= \mathbf{D}\frac{b}{x+b} + \frac{x}{x+b}\mathbf{C} \\
&= \left[\mathbf{A} + \left(\frac{b}{a+b}\right)(\mathbf{B} - \mathbf{A})\right]\frac{b}{x+b} + \frac{x}{x+b}\mathbf{C} \qquad \text{(substituting for } \mathbf{D}\text{)}.
\end{aligned}$$

Substituting

$$x = \frac{bc}{a+b},$$

we find that

$$\frac{b}{x+b} = \frac{b}{\frac{bc}{a+b}+b} = \frac{a+b}{c+a+b} \quad \text{and} \quad \frac{x}{x+b} = \frac{\frac{bc}{a+b}}{\frac{bc}{a+b}+b} = \frac{c}{c+a+b}.$$

Hence

$$I = \left[\mathbf{A} + \left(\frac{b}{a+b} \right) (\mathbf{B} - \mathbf{A}) \right] \frac{a+b}{c+a+b} + \frac{c}{c+a+b} \mathbf{C}$$

$$= \frac{1}{a+b+c} [(a+b)\mathbf{A} + b\mathbf{B} - b\mathbf{A}] + \frac{c}{c+a+b} \mathbf{C}$$

$$= \frac{1}{a+b+c} (a\mathbf{A} + b\mathbf{B} + c\mathbf{C}),$$

that is

$$\mathbf{I} = \frac{a\mathbf{A} + b\mathbf{B} + c\mathbf{C}}{p},$$

where p is the perimeter of the triangle.

8 A Nasty Integral

(Problem 997, Fall 2001; posed by Robert C. Gebhardt, Hopatcong, New Jersey)

Solution by Sophie Trawalter, Student, University of North Carolina at Wilmington.

Evaluate the integral

$$I = \int_4^8 \frac{\ln(9-x)\,dx}{\ln(9-x) + \ln(x-3)}$$

First make the substitution $x = y + 6$ to give

$$I = \int_{-2}^2 \frac{\ln(3-y)\,dy}{\ln(3-y) + \ln(3+y)} \tag{1}$$

Now substitute $y = -u$ to give

$$I = \int_2^{-2} \frac{-\ln(3+u)\,du}{\ln(3+u) + \ln(3-u)}$$

Interchanging the limits cancels the minus sign in the numerator to give

$$I = \int_{-2}^{2} \frac{\ln(3+u)\,du}{\ln(3+u) + \ln(3-u)}$$

which, upon changing the variable of integration back to y, yields

$$I = \int_{-2}^{2} \frac{\ln(3+y)dy}{\ln(3+y) + \ln(3-y)} \tag{2}$$

Adding (1) and (2), we get

$$2I = \int_{-2}^{2} \frac{\ln(3-y) + \ln(3+y)}{\ln(3+y) + \ln(3-y)}\,dy$$

$$= \int_{-2}^{2} dy$$

$$= 4.$$

Hence $I = 2$.

Congratulations Sophie!

9 Another Delightful Surprise From Leon Bankoff

(Problem 1041, Spring 2002)

In the quarter circle in Figure 8, prove that the three white circles are the same size and also that the six smaller shaded circles are the same size.

Leon has taken it easy on us this time.

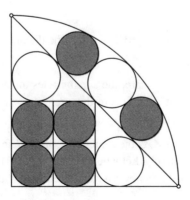

FIGURE 8

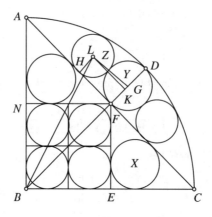

FIGURE 9

Clearly, in Figure 9, the circle Y meets the arc at its midpoint D, making F the midpoint of AC and E the midpoint of BC. Hence, let $BC = 2$. Then $\triangle ECF$ is an isosceles right triangle with sides $1, 1, \sqrt{2}$.

Now, it is fairly well known that

> the diameter of the incircle of a right triangle
> = the sum of the legs minus the hypotenuse.

(This is proved in the Appendix to this section.) Hence the diameter of the circle X is $2 - \sqrt{2}$.

But diagonal BF of unit square $BEFN$ is also $\sqrt{2}$, and so the equation

(the diameter FD of Y) + BF = radius 2 of the quarter-circle,

gives

the diameter of $Y = 2 - \sqrt{2} =$ the diameter of X,

making X and Y the same size.

Since $BEFN$ is a unit square, the small circles inside it have radius $\frac{1}{4}$. It remains only to show that the radius of the circle Z is $\frac{1}{4}$; the symmetry of the figure takes care of everything else.

Let L be the center of Z, G the center of Y, and LH and LK be perpendicular to AC and BD, respectively. Let r be the radius of Z. Then $LH = FK = r$.

In right triangle LGK, then, we have, since Z and Y are tangent,

$$LG = \text{the sum of the radii} = r + \left(1 - \tfrac{\sqrt{2}}{2}\right).$$

Also

$$KG = FG - FK = \text{the difference of these radii } = 1 - \tfrac{\sqrt{2}}{2} - r.$$

By the theorem of Pythagoras, then, we have

$$LK^2 = LG^2 - KG^2 = \left(1 - \tfrac{\sqrt{2}}{2} + r\right)^2 - \left(1 - \tfrac{\sqrt{2}}{2} - r\right)^2$$
$$= \left(2 - \sqrt{2}\right)(2r) = 4r - 2\sqrt{2}r.$$

Finally, in right triangle LBK, we have

$$BL^2 = BK^2 + LK^2$$
$$= \left(\sqrt{2} + r\right)^2 + 4r - 2\sqrt{2}r$$
$$= r^2 + 4r + 2. \tag{1}$$

Now, since Z and the quarter circle are tangent,

$$BL = \text{the difference between their radii } = 2 - r.$$

Hence

$$BL^2 = 4 - 4r + r^2 = r^2 + 4r + 2 \quad \text{(from (1))},$$

giving $2 = 8r$, and the desired $r = \tfrac{1}{4}$.

10 A Problem of Professor M. V. Subbarao

(Problem 1020, Spring 2001)

Let $p_1, p_2, \ldots, p_r$ be r distinct odd primes, and let a be any fixed integer. You are given that

$$(p_1 + a)(p_2 + a)\ldots(p_r + a) - 1$$

is divisible by

$$(p_1 + a - 1)(p_2 + a - 1)\ldots(p_r + a - 1).$$

For $r = 1$, this is trivial, namely that $p_1 + a - 1$ is divisible by $p_1 + a - 1$. Are there primes and an integer a such that this holds for an $r > 1$?

For $a = 0$, this is an unsolved problem that was posed in 1932 by the renowned number theorist D.H. Lehmer, and for $a = 1$, it is one of Professor Subbarao's own unsolved problems, posed as recently as 1998. Professor Subbarao is also a distinguished number theorist. He dedicated this problem to the well known problem solver and creator Murray Klamkin, who has been his long-time friend and colleague at the University of Alberta in Edmonton. In fact, Professor Subbarao's enthusiasm for the occasion took substance in his offer of \$100 for the first valid solution that was received by the Problems Section of this journal.

Here is the marvelous "twin prime" solution of Chetan Offord and Robert Wentz of St. John's University, Collegeville, Minnesota.

For $r = 2$, one need only let p_1 and $p_2 = p_1 + 2$ be any twin primes, and take $a = -p_1$. Then

$$(p_1 + a)(p_2 + a) - 1 = 0(p_2 + a) - 1 = -1$$

and

$$(p_1 + a - 1)(p_2 + a - 1) = -1(p_2 - p_1 - 1) = -1(2 - 1) = -1.$$

Needless to say, Offord and Wentz got the \$100.

Appendix

Let triangle ABC be right-angled at C. Then, its indiameter d equals the sum of the legs minus the hypotenuse: $d = a + b - c$.

Proof: From $\Delta = rs$ and $a^2 + b^2 = c^2$, we have

$$d = 2r = \frac{2\Delta}{s} = \frac{4\Delta}{2s} = \frac{2ab}{a+b+c}$$
$$= \frac{2ab(a+b-c)}{(a+b)^2 - c^2} = \frac{2ab(a+b-c)}{2ab} = a + b - c.$$

From *Problems in Plane Geometry*

Viktor Prasolov's *Problems in Plane Geometry* is a monumental collection of some 2000 problems that is encyclopedic in its coverage. It is a goldmine of exciting problems and brilliant solutions. The following gems from just the first two of its thirty chapters don't even scratch the surface of this amazing book.

1 From Chapter One: Similar Triangles

1. (#42) *ABC* is an isosceles right angled triangle with the right angle at *C* (Figure 1). Points *D* and *E*, equidistant from *C*, are chosen arbitrarily on *AC* and *BC*. Perpendiculars from *D* and *C* to *AE* meet the hypotenuse *AB* at *K* and *L*. Prove *L* is the midpoint of *KB*.

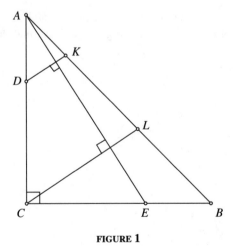

FIGURE 1

A clockwise quarter turn about *C* takes △*ACE* to △*BCF*, where *F* is on *AC* extended and *DC* = *CE* = *CF* (Figure 2). This also rotates *AE*

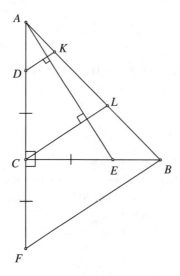

FIGURE 2

through a right angle, making its image FB parallel to CL and DK. Since C is the midpoint of DF, then CL is the midline of the strip between parallels DK and FB, and hence it bisects every transversal across the strip, in particular KB.

2. (#60) Tangents to the incircle of ABC are drawn parallel to the sides to cut off a little triangle at each vertex (Figure 3). Prove that the inradii of the

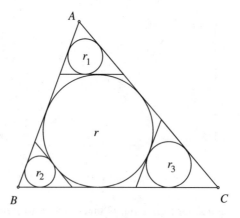

FIGURE 3

three small triangles sum to the inradius of $\triangle ABC$:

$$r_1 + r_2 + r_3 = r.$$

A half-turn about the incenter of $\triangle ABC$ takes tangents PQ and PU along their parallel tangents TS and SR, and thus takes P to S (Figure 4). Similarly Q goes to T and we have $PQ = TS$. Thus side

$$BC = BT + TS + SC = BT + PQ + SC.$$

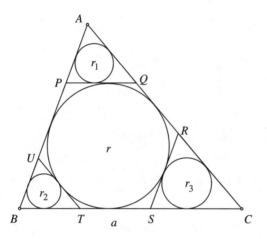

FIGURE 4

Now, $\triangle APQ$ is similar to $\triangle ABC$, and so

$$\frac{PQ}{BC} = \frac{r_1}{r},$$

giving

$$PQ = \frac{r_1}{r}a.$$

Similarly

$$BT = \frac{r_2}{r}a, \quad \text{and} \quad SC = \frac{r_3}{r}a,$$

and we have

$$BC = \frac{r_1}{r}a + \frac{r_2}{r}a + \frac{r_3}{r}a = a,$$

which easily reduces to $r_1 + r_2 + r_3 = r$.

Now for a remarkable property.

3. (#13) *AD* and *BE* are angle bisectors in $\triangle ABC$ (Figure 5). Prove that the distance to *AB* from any point *M* on *DE* is the sum of its distances to *BC* and *AC*:

$$z = x + y.$$

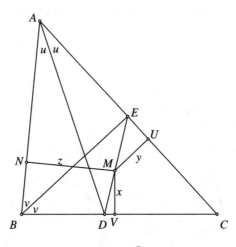

The proof is a simple application of the following lemma.

Lemma. *If ABCD is a trapezoid with parallel sides AD = b and BC = a, (Figure 6), then the length of the segment MN which joins the points M and N that divide each of AB and CD in the ratio p : q is*

$$MN = \frac{ap + bq}{p + q}.$$

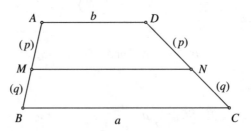

FIGURE 6

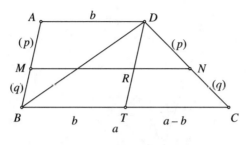

FIGURE 7

Proof: First let us show that *MN* is parallel to *AD* and *BC* (Figure 7):

Since a line across a triangle that is parallel to a side of the triangle divides the other two sides proportionally, in $\triangle ABD$, the line through *M* which is parallel to *AD* would divide *DB* in the ratio $p : q$, and then, since it would also be parallel to *BC*, would continue across $\triangle DBC$ to divide *DC* in the same ratio, and would therefore meet *DC* in *N*.

Suppose $a > b$. Let *DT* be drawn across the trapezoid to complete parallelogram *ABTD*. Let *DT* cross *MN* at *R*. Then *MR* and *BT* are both of length *b*, and $TC = a - b$ (Figure 7).

In $\triangle DTC$, *RN* is parallel to *TC*, and so

$$\frac{RN}{TC} = \frac{p}{p+q}, \quad \text{giving } RN = (a - b) \cdot \frac{p}{p+q}.$$

Hence

$$MN = MR + RN$$
$$= b + (a - b) \cdot \frac{p}{p+q}$$
$$= \frac{bp + bq + ap - bp}{p+q},$$

and

$$MN = \frac{ap + bq}{p+q}.$$

Since this formula is trivially valid for $a = b$, that is, for a parallelogram *ABCD*, it is clear that it doesn't matter whether *a* is bigger, smaller, or the same length as *b*, and the proof of the lemma is complete.

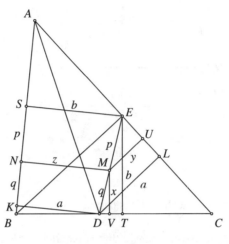

FIGURE 8

Returning to the problem, suppose M divides ED in the ratio $p : q$ and that ES and DK are perpendicular to AB (Figure 8). Since MN is parallel to these perpendiculars, N divides SK in the ratio $p : q$. Let $SE = b$ and $DK = a$. Similarly, let DL and ET be perpendiculars to AC and BC. Then, since D and E are on the angle bisectors of angles A and B, we have $DL = DK = a$ and $ET = ES = b$. Now, $SKDE$ is a trapezoid in which M and N divide the non-parallel sides in the ratio $p : q$. By the lemma, then, the length z of MN is

$$z = \frac{ap + bq}{p + q}.$$

In $\triangle DET$, MV is parallel to ET, and so

$$\frac{x}{b} = \frac{q}{p + q}, \quad \text{giving } x = \frac{bq}{p + q}.$$

Similarly, MU is parallel to DL and

$$\frac{y}{a} = \frac{p}{p + q}, \quad \text{giving } y = \frac{ap}{p + q},$$

and it is clear that $x + y = z$, as desired.

4. (#14) Side CD of rectangle $ABCD$ is extended an arbitrary distance beyond D to a point P (Figure 9). The line joining P to the midpoint M of AD meets diagonal AC at Q. If N is the midpoint of BC, prove that MN bisects $\angle QNP$.

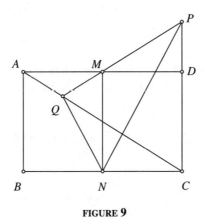

Clearly the midline *MN* and diagonal *AC* meet at the center *O* of the rectangle, and *O* is the midpoint of *MN* (Figure 10). Let the perpendicular to *MN* through *O* meet *QN* at *K*. Then *OK* is the perpendicular bisector of *MN* and *KM* = *KN*. Then △*KMN* is isosceles with equal base angles *z* at *M* and *N*.

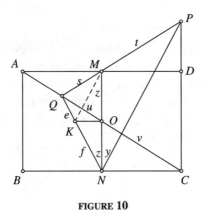

FIGURE 10

Then, in △*PQC*, *MO* is parallel to *PC*, implying

$$\frac{s}{t} = \frac{u}{v};$$

and in △*QNC*, *KO* is parallel to *NC*, implying

$$\frac{u}{v} = \frac{e}{f}.$$

Hence

$$\frac{s}{t} = \frac{e}{f},$$

from which it follows that *KM* is parallel to *NP*. Thus alternate angles *y* and *z* at *N* and *M* are equal and the conclusion follows.

5. (#57) *AD*, *BE*, and *CF* are the altitudes of acute triangle *ABC* (Figure 11). Prove that reflection in *AC* takes *D* to a point *G* on *FE* extended.

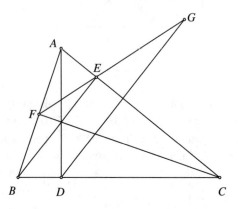

FIGURE 11

Referring to Figure 12, quadrilateral *FBCE* is cyclic because *BC* subtends a right angle at each of *E* and *F*, and so the exterior angle ∠*AEF* is equal to the interior ∠*B* at the opposite vertex. Similarly, *ABDE* is cyclic and exterior ∠*DEC* equals ∠*B* at the opposite vertex. Since the reflection takes ∠*DEC* to ∠*CEG*, the angle *B* occurs three times at *E* as shown. Thus ∠*FEG* is equal to straight angle *AEC*.

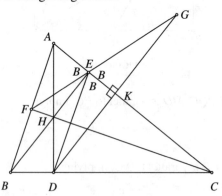

FIGURE 12

2 From Chapter 2: Inscribed Angles

1. (#48) (Another surprise)

   AD, *BE*, and *CF* are the altitudes of $\triangle ABC$ (Figure 13). From an arbitrary point *K* on *BC*, a line is drawn parallel to *CF* to meet *EF* extended at *L*. For all points *K* on *BC*, prove that the circumenter of $\triangle LDK$ always lies on *AC*.

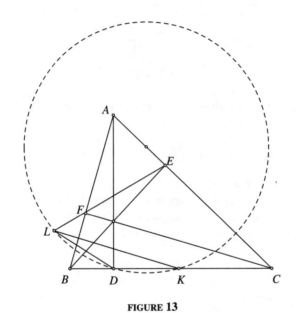

FIGURE 13

Since *AC* subtends right angles at *F* and *D*, *AFDC* is cyclic (Figure 14); moreover, *AC* is a **diameter** of the circumcircle. Now, reflecting a point on a circle in a diameter always gives another point on the circle; in particular, reflecting *D* in *AC* gives another point *G* on the circle around *AFDC*.

By the immediately preceding problem (chapter 1, #57), we showed that this image *G* lies on *FE* extended. Hence, in the circle *AFDC*, chord *FD* subtends equal angles *x* at *G* and *C*. But *LK* is parallel to *FC*, and hence $\angle LKD = \angle FCD = x$. Thus *LD* subtends equal angles *x* at *K* and *G*, implying that *LDKG* is cyclic.

In the circle around *LDKG*, which is clearly the circumcircle of $\triangle KLD$, *DG* is a chord, and therefore its perpendicular bisector, namely

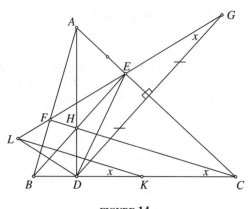

FIGURE 14

the line AC, passes through the center of the circle (recall that G is the reflection of D in AC).

2. (#12 and its introductory problem, #11.)

Suppose a convex $2n$-gon $A_1 A_2 \ldots A_{2n}$ is inscribed in a circle so that every pair of opposite sides is parallel except one. Prove that (a) if n is odd, the remaining two sides are also parallel (Figure 15), (b) if n is even, the remaining two sides are equal (Figure 16).

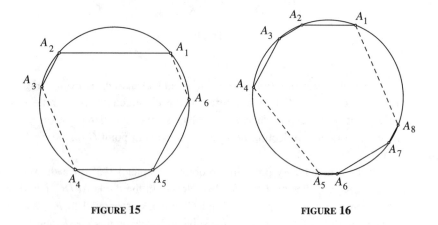

FIGURE 15   **FIGURE 16**

(i) Consider first the case of a hexagon, i.e., $n = 3$, where (AB, DE) and (BC, EF) are pairs of parallel sides (Figure 17). Let us show that CD and AF are also parallel.

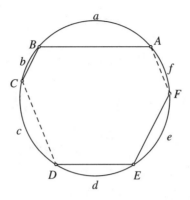

We observe that, since reflection in the common perpendicular bisector of two parallel chords in a circle interchanges the arcs that lie between the two chords, it follows that the arcs are equal; for example, in Figure 17, arcs *BCD* and *EFA* lie between the parallel chords *AB* and *DE* and would reflect into one another in the common perpendicular bisector of *AB* and *DE*. Conversely, chords which join the appropriate ends of equal arcs are parallel.

Hence, labelling the arcs as shown in Figure 17, we have that the parallel sides *AB* and *DE* give $b + c = e + f$, and the parallel sides *BC* and *EF* give $c + d = f + a$. Subtracting gives $b - d = e - a$ and therefore $a + b = d + e$, implying that the sides *CD* and *FA* are parallel.

Now consider the general case of $2n$-gon $A_1 A_2 \ldots A_n B_1 B_2 \ldots B_n$ in which all the opposite pairs are given parallel except the pair $(A_n B_1, B_n A_1)$.

(a) By part (i), in hexagon $A_1 A_2 A_3 B_1 B_2 B_3$ (Figure 18), the sides $A_3 B_1$ and $A_1 B_3$ are parallel. More to the point,

$$\text{arc } A_1 A_3 = \text{arc } B_1 B_3.$$

Similarly, from $A_3 A_4 A_5 B_3 B_4 B_5$, we have

$$\text{arc } A_3 A_5 = \text{arc } B_3 B_5,$$

and altogether, then, arc $A_1 A_5 = $ arc $B_1 B_5$. Continuing around the polygon, we obtain

$$\text{arc } A_1 A_{2k+1} = \text{arc } B_1 B_{2k+1}.$$

Thus, for odd $n (= 2k + 1)$, the final pair of sides, $A_n B_1$ and $B_n A_1$, lie between equal arcs and are therefore parallel.

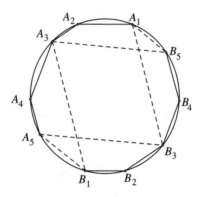

FIGURE 18

(b) Suppose n is even, say $n = 2k$, and all $2k$ pairs of the polygon $A_1 A_2 \ldots A_{2k} B_1 B_2 \ldots B_{2k}$ are given parallel except $A_{2k} B_1$ and $B_{2k} A_1$ (Figure 19). Thus, momentarily neglecting the vertices A_{2k} and B_{2k}, in the remaining $(4k-2)$-gon $A_1 A_2 \ldots A_{2k-1} B_1 B_2 \ldots B_{2k-1}$, all the pairs of opposite sides are parallel except the pair $A_{2k-1} B_1$ and $B_{2k-1} A_1$. By the result just proved for n odd, it follows that these diagonals are also parallel. Now, the sides $A_{2k-1} A_{2k}$ and $B_{2k-1} B_{2k}$ are given parallel, and so the respective angles x and y between these parallel sides and the parallel diagonals $A_{2k-1} B_1$ and $B_{2k-1} A_1$ are equal (Figure 19):

$$x = \angle A_{2k} A_{2k-1} B_1 = \angle A_1 B_{2k-1} B_{2k} = y.$$

Thus the final pair of opposite sides, $A_{2k} B_1$ and $B_{2k} A_1$, subtend equal angles at the circumference and are therefore equal in length.

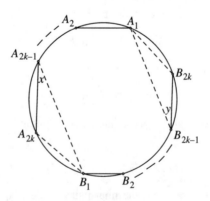

FIGURE 19

3. (#9) A circle is divided into equal arcs by n diameters (Figure 20). Prove that the feet of the perpendiculars to these diameters from a point P inside the circle always determine a regular n-gon N.

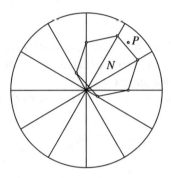

FIGURE 20

Let A, B, C be three consecutive vertices of N (Figure 21); let O be the center of the circle and x the angle between consecutive diameters.

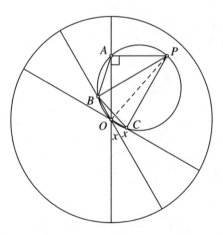

FIGURE 21

Clearly, the circle on diameter OP goes through each vertex of N. Thus AB subtends equal angles at P and O on this circle and we have

$$\angle APB = \angle AOB = x.$$

Now, *BOCP* is cyclic and therefore the interior angle $\angle BPC$ is equal to the exterior angle x at the opposite vertex O. Thus chords AB and BC subtend the same angle x at P and are therefore equal. Similarly, it follows that all the sides of N are the same length and, since N is cyclic, that N is indeed a regular n-gon.

4. (#68) On side AB of $\triangle ABC$ an arbitrary point P is taken (Figure 22). PN and PM are drawn respectively parallel to BC and AC. The circumcircles of triangles APN and PBM intersect at Q.

   Prove that, as P varies on AB, all the lines PQ are concurrent.

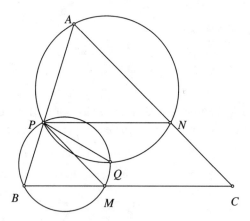

FIGURE 22

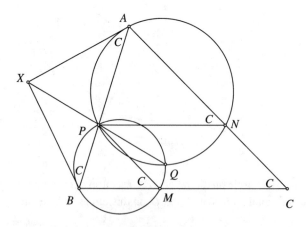

FIGURE 23

Construct $\angle C$ at A and B to give isosceles triangle XAB outwardly on AB (Figure 23); then $XA = XB$.

Now, the parallel lines make $\angle C$ equal to both $\angle ANP$ and $\angle PMB$. Thus XA and XB are equal **tangents** to the circles, implying X is always a point on their radical axis which, in the case of these intersecting circles, is their common chord PQ. Thus, as P varies on AB, PQ always goes through the fixed point X. (A brief discussion of radical axes is given in the Appendix to Section 4.)

From *The New Mexico Mathematics Contest Problem Book*

This excellent book by Liong-shin Hahn is soon to be published by the University of New Mexico Press. The book contains a wealth of ingenious poblems, many of Professor Hahn's own devising, and it is most warmly recommended.

1. Find the minimum term in the sequence

$$\sqrt{\frac{7}{6}} + \sqrt{\frac{96}{7}}, \sqrt{\frac{8}{6}} + \sqrt{\frac{96}{8}}, \sqrt{\frac{9}{6}} + \sqrt{\frac{96}{9}}, \ldots, \sqrt{\frac{95}{6}} + \sqrt{\frac{96}{95}}.$$

By the A.M.–G.M. inequality we have

$$\sqrt{\frac{n}{6}} + \sqrt{\frac{96}{n}} \geq 2\sqrt{\sqrt{\frac{n}{6}} \cdot \sqrt{\frac{96}{n}}} = 2\sqrt{4} = 4,$$

with equality if and only if

$$\sqrt{\frac{n}{6}} = \sqrt{\frac{96}{n}},$$

i.e., for $n = \sqrt{576} = 24$.

Hence the minimum term is

$$\sqrt{\frac{24}{6}} + \sqrt{\frac{96}{24}} = 4.$$

2. (a) Find positive integers a and b such that

$$\tan \frac{3\pi}{8} = a + \sqrt{b}.$$

(b) Show that

$$\left(\tan \frac{3\pi}{8} \right)^n + (-1)^n \left(\cot \frac{3\pi}{8} \right)^n$$

is an even integer for every positive integer n.

(c) For each positive integer n, let

$$k_n = \left[\left(\tan\frac{3\pi}{8}\right)^n\right],$$

the integer part of

$$\left(\tan\frac{3\pi}{8}\right)^n.$$

Show that k_n and n are always of opposite parity.

(a) First of all, the value of $\tan(3\pi/8)$ can be determined very cleverly as indicated in Figure 1 (recall the first "Mathematics without words" in part 5 of Section 5):

First, an isosceles right triangle $\triangle ABC$ of leg 1 is constructed. Thus each base angle $x = \pi/4$ and the length of the hypotenuse AC is $\sqrt{2}$. Next let AC be unfolded along BC extended to give $CD = \sqrt{2}$. Thus $\triangle ACD$ is also isosceles and its base angles y add up to the exterior angle x at C, making

$$y = \frac{1}{2}\cdot\frac{\pi}{4}.$$

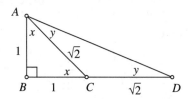

FIGURE 1

Now

$$\frac{3\pi}{8} = \frac{2\pi}{8} + \frac{\pi}{8} = \frac{\pi}{4} + \frac{1}{2}\cdot\frac{\pi}{4} = x + y = \angle BAD,$$

and so

$$\tan\frac{3\pi}{8} = \frac{BD}{AB} = 1 + \sqrt{2}.$$

(b) It follows that

$$\cot\frac{3\pi}{8} = \frac{1}{1+\sqrt{2}}\cdot\frac{1-\sqrt{2}}{1-\sqrt{2}} = -(1-\sqrt{2}),$$

making

$$\left(\tan\frac{3\pi}{8}\right)^n + (-1)^n \left(\cot\frac{3\pi}{8}\right)^n = \left(1+\sqrt{2}\right)^n + (-1)^{2n}\left(1-\sqrt{2}\right)^n$$
$$= \left(1+\sqrt{2}\right)^n + \left(1-\sqrt{2}\right)^n.$$

Expanding by the binomial theorem yields an expression in which the irrational terms cancel and the integer terms double up, resulting in an even integer.

(c) From part (b) we have $(1+\sqrt{2})^n + (1-\sqrt{2})^n = 2t$ for some integer t. Hence

$$\left(\tan\frac{3\pi}{8}\right)^n = \left(1+\sqrt{2}\right)^n = 2t - \left(1-\sqrt{2}\right)^n \dots \tag{1}$$

Now, $(1-\sqrt{2})$ is negative, and so when n is odd, $-(1-\sqrt{2})^n$ is positive, and when n is even, $-(1-\sqrt{2})^n$ is negative. Since the magnitude of $1-\sqrt{2}$ is less than 1, so is the magnitude of $(1-\sqrt{2})^n$. Hence, for all n,

$$\left(\tan\frac{3\pi}{8}\right)^n$$

differs from the even number $2t$ by less than unity. From (1), then, it follows that, for n odd,

$$\left(\tan\frac{3\pi}{8}\right)^n$$

exceeds $2t$, making

$$k_n = \left[\left(\tan\frac{3\pi}{8}\right)^n\right] = 2t,$$

an even number, and for n even,

$$k_n = 2t - 1,$$

an odd number. The conclusion follows.

3. Find a necessary and sufficient condition on the parameter k so that the equation

$$2\sin\theta + k\cos\theta = \sqrt{7}$$

has a solution.

The key is to write the left side in a way that may be construed as the sine of the sum of two angles. To this end,

$$2\sin\theta + k\cos\theta = \sqrt{2^2 + k^2}\left(\frac{2}{\sqrt{2^2+k^2}}\sin\theta + \frac{k}{\sqrt{2^2+k^2}}\cos\theta\right).$$

Now,

$$\frac{2}{\sqrt{2^2+k^2}} \le 1,$$

and setting

$$\frac{2}{\sqrt{2^2+k^2}} = \cos\varphi,$$

it follows that

$$\sin\varphi = \sqrt{1 - \cos^2\varphi} = \frac{k}{\sqrt{2^2+k^2}},$$

and

$$2\sin\theta + k\cos\theta = \sqrt{2^2+k^2}\sin(\theta + \varphi).$$

Thus the given equation has a solution if and only if there is a solution to

$$\sqrt{2^2+k^2}\sin(\theta+\varphi) = \sqrt{7},$$

that is,

$$\sin(\theta+\varphi) = \sqrt{\frac{7}{2^2+k^2}}.$$

Since the sine always has magnitude ≤ 1, this is equivalent to

$$\frac{7}{2^2+k^2} \le 1, \quad \text{and} \quad |k| \ge \sqrt{3}.$$

4. What is the sum of the twelve angles marked o in Figure 2?

In Figure 3, the angles at P and Q add up to the exterior angle a; similarly the angles at R and S add to b, and the remaining pairs add to exterior angles c, d, e, and f.

Now, in $\triangle UVW$, $(g+h)+i = 180°$, while along SQ,

$$(g+h)+a+b = 2\cdot 180°.$$

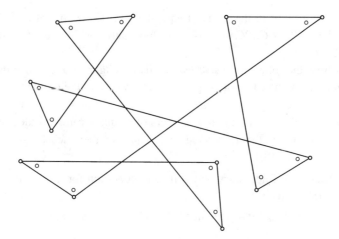

FIGURE 2

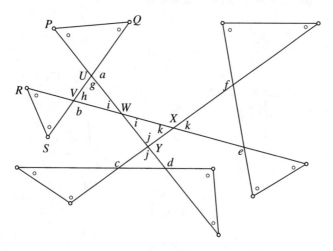

FIGURE 3

Hence, $a+b = i+180°$. Similarly, $c+d = j+180°$ and $e+f = k+180°$. Altogether, then, since $i + j + k = 180°$ in $\triangle WXY$, the twelve angles add to

$$a + b + c + d + e + f = (i + j + k) + 3 \cdot 180°$$
$$= 4 \cdot 180°$$
$$= 720°.$$

5. Let P be a point in the plane of equilateral $\triangle ABC$ such that each of the triangles PAB, PBC, PCA is isosceles. How many positions are there for the point P?

    Since the solution consists essentially of a figure, it is given at the end of this section in order not to spoil your enjoyment of the problem.

6. Clearly, in Figure 4(a), the four corners of a square can be folded over to meet at a point without overlapping or gaps; another such figure is illustrated in Figure 4(b).

    (a) Determine a necessary and sufficient condition for a quadrilateral to permit such a folding.

    (b) In Figure 5, the inscribed quadrilateral $ABCD$ is such that

    $$\text{arc } AB + \text{arc } CD = \text{arc } BC + \text{arc } AD.$$

    Determine whether $ABCD$ allows such a folding.

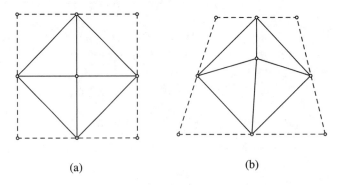

(a)   (b)

FIGURE 4

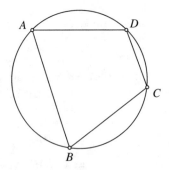

FIGURE 5

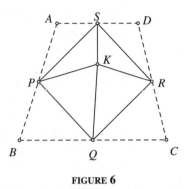

FIGURE 6

(a) Suppose that folding A about PS, B about PQ, C about QR, and D about RS, carries each vertex to K without overlapping or gaps (Figure 6). Thus K is the reflection of A in PS and also the reflection of B in PQ. In these reflections, both AP and BP are taken to PK, implying that $AP = BP$. Thus P must be the midpoint of AB; similarly Q, R, and S must be the midpoints of the other sides.

Now, A and its reflection K are the same distance from PS. But, in $\triangle ABD$, since P and S are the midpoints of their sides, then PS is halfway between A and BD, and so K must lie on BD. Moreover, since K is the image of A in PS, AK is perpendicular to PS, and since PS is parallel to BD in $\triangle ABD$, it follows that K must be the foot of the altitude from A in $\triangle ABD$.

For a folding, then, this image K must be the foot of the altitude from each of A, B, C, D, in their respective triangles. It follows, then, that the diagonals AC and BD must intersect at right angles at K. Hence a necessary condition for a folding is that the diagonals be perpendicular.

It is easy to see that this is also a sufficient condition: if the diagonals are perpendicular (Figure 7), their point of intersection K is the foot of

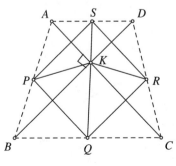

FIGURE 7

the altitude from A in $\triangle ABD$, the foot of the altitude from B in $\triangle ABC$, and so on; reflecting A in the midline PS, which is parallel to diagonal BD and bisects AK, would take it to K; also, AP would be carried to PK, as would BP when B is reflected in the midline PQ; and so on around the quadrilateral. Clearly this gives a folding without overlapping or gaps.

(b) Now let us consider the quadrilateral in Figure 8. Since the arcs AB and CD add to a semicircle, the sum of the angles these arcs subtend at the center is a straight angle. At the circumference, then, the sum of the angles x and y subtended by these arcs is a right angle, and in triangle ADK, angle K is a right angle, making the diagonals perpendicular and implying that $ABCD$ does indeed allow a folding.

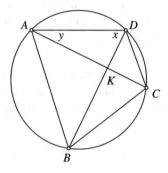

FIGURE 8

7. In Figure 9, P is a variable point in the interior of segment AB, and triangles QAP and RPB are equilateral. Determine the locus of the point of intersection K of AR and BQ.

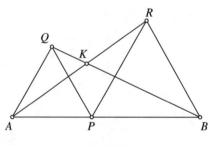

FIGURE 9

The brilliant idea here is to consider the 60° clockwise rotation with center P. This takes A to Q and R to B, thus taking $\triangle ARP$ to $\triangle QBP$. Hence $\angle ARP = \angle QBP$.

Thus KP subtends the same angle at R and B, implying $KPBR$ is cyclic. Therefore RB subtends a 60° angle at K as it does at P. Hence $\angle AKB$ is always 120°, implying that the locus of K is a 120°-arc of a circle on chord AB.

8. In Figure 10, AB is a diameter, $AC = DC = 3$, and $DB = 7$. What is the radius of the circle?

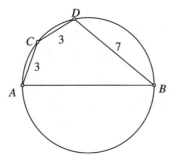

FIGURE 10

Again a brilliant solution. Consider the reflection in the perpendicular bisector of CB (reflecting a circle in the perpendicular bisector of a chord takes the circle into itself). This interchanges C and B and takes CD to a chord $BE = CD = 3$ (Figure 11). Since $BE = AC$, then CE is parallel to AB. This reflection also takes BD to CE, and so $CE = BD = 7$.

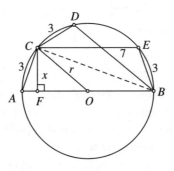

FIGURE 11

Let $CF = x$ be the perpendicular from C to AB, and let the radius be r. Now, $ABEC$ is an isosceles trapezoid, in which case $CE = 2.OF$ (projecting CE onto AB would give a segment whose midpoint is O). Hence $OF = \frac{7}{2}$, implying $AF = r - \frac{7}{2}$.

In right triangle CFO, then, we have

$$x^2 = r^2 - OF^2 = r^2 - \frac{49}{4},$$

and from right triangle CAF, we have

$$x^2 = 9 - AF^2 = 9 - \left(r - \frac{7}{2}\right)^2.$$

Hence

$$r^2 - \frac{49}{4} = 9 - \left(r - \frac{7}{2}\right)^2,$$
$$r^2 - \frac{49}{4} = 9 - r^2 + 7r - \frac{49}{4},$$
$$2r^2 - 7r - 9 = 0,$$
$$(2r - 9)(r + 1) = 0,$$

and since r is positive, $r = \frac{9}{2}$.

9. Finally, let us close with the following intriguing problem.

Does there exist a finite sequence of numbers $\{a_k\}$, real or complex, such that $|a_k| < 1$ for all k, and that, summed over all terms,

$$\left|\sum a_k\right| < \left|\sum a_k^2\right| < \left|\sum a_k^3\right| < \left|\sum a_k^4\right| < \left|\sum a_k^5\right|?$$

If the chain of inequalities had only two links,

$$\left|\sum a_k\right| < \left|\sum a_k^2\right|,$$

an obvious solution would be $a_1 = r$, $a_2 = -r$, where $0 < r < 1$: then

$$|a_k| = r < 1, \quad \left|\sum a_k\right| = 0, \quad \text{and} \quad \left|\sum a_k^2\right| = 2r^2 > 0.$$

Encouraged, let us see if we can take on the three-link chain.

There is a clue in this solution of the two-link chain, if we could only see it. Eventually it might dawn on us that this solution is $\{r \cdot 1, r(-1)\}$ and that 1 and -1 are the two square roots of unity. There might be a chance, then, that r times the three cube roots of unity figure somehow in the solution of the three-link chain.

The cube roots of unity are 1, ω, and ω^2, where

$$\omega = \frac{-1 + i\sqrt{3}}{2}.$$

Being the roots of $x^3 - 1 = 0$, it follows that

$$\text{the sum of the roots} = 1 + \omega + \omega^2 = 0.$$

Moreover, since $\omega^3 = 1$, the sum of their squares is

$$1 + \omega^2 + \omega^4 = 1 + \omega^2 + \omega = 0.$$

Things are not looking good with both $\left| \sum a_k \right|$ and $\left| \sum a_k^2 \right|$ equal to 0. Thus the sequence $\{r, r\omega, r\omega^2\}$, on its own, doesn't even get us as far as the second link.

However, in a transport of inspiration, let's include the r and $-r$ of the two-link chain to give a 5-term sequence

$$\{r, -r, r, r\omega, r\omega^2\}.$$

Then

$$\left| \sum a_k \right| = \left| r - r + r(1 + \omega + \omega^2) \right| = \left| 0 + r(0) \right| = 0,$$

$$\left| \sum a_k^2 \right| = \left| r^2 + r^2 + r^2(1 + \omega^2 + \omega^4) \right| = \left| 2r^2 + r^2(0) \right| = 2r^2,$$

$$\left| \sum a_k^3 \right| = \left| r^3 - r^3 + r^3(1 + \omega^3 + \omega^6) \right| = \left| 0 + 3r^3 \right| = 3r^3.$$

Thus, choosing r so that $\frac{2}{3} < r < 1$, makes both $2r^2 < 3r^3$ and $|a_k| = r < 1$ for all k.

It seems that we might be on to something here. Let's check whether a solution to the four-link chain is given by the nine-term sequence whose terms are r times the square, cube, and fourth roots of unity.

The fourth roots of unity are 1, -1, i, $-i$, where $i = \sqrt{-1}$, making the proposed sequence $\{r, -r, r, r\omega, r\omega^2, r, -r, ri, -ri\}$. Clearly

$$\left| \sum a_k \right| = 0,$$

$$\left| \sum a_k^2 \right| = \left| r^2 + r^2 + r^2(1 + \omega^2 + \omega^4) + r^2(1 + 1 - 1 - 1) \right| = 2r^2,$$

$$\left| \sum a_k^3 \right| = \left| r^3 - r^3 + r^3(1 + \omega^3 + \omega^6) + r^3(1 - 1 - i + i) \right| = 3r^3,$$

and

$$\left| \sum a_k^4 \right| = \left| r^4 + r^4 + r^4(1 + \omega^4 + \omega^8) + r^4(1 + 1 + 1 + 1) \right| = 6r^4.$$

(Observe that $1 + \omega^4 + \omega^8 = 0$.)

It remains only to see whether r can be chosen to satisfy $|r| < 1$ and the three inequalities $0 < 2r^2 < 3r^3 < 6r^4$.

The first inequality only requires $0 < r < 1$, the second that $\frac{2}{3} < r < 1$, the third that $\frac{1}{2} < r < 1$. Thus any r in the range $\frac{2}{3} < r < 1$ gives a solution to the four-link chain.

We have every confidence, then, that adding to this solution r times the fifth roots of unity will provide a solution to the five-link chain. However, we had better make sure.

In order to simplify the calculations, we might observe that the nth roots of unity may be denoted by

$$1, \omega_n, \omega_n^2, \omega_n^3, \ldots, \omega_n^{n-1},$$

where $\omega_n = e^{2\pi i/n}$. Thus $\omega_n^n = 1$ and the sum of their mth powers is the n-term geometric series

$$1 + \omega_n^m + \omega_n^{2m} + \omega_n^{3m} + \cdots + \omega_n^{(n-1)m} = \frac{1 - (\omega_n^m)^n}{1 - \omega_n^m}$$

$$= \frac{1 - (\omega_n^n)^m}{1 - \omega_n^m} = \frac{1 - 1}{1 - \omega_n^m} = \frac{0}{1 - \omega_n^m}.$$

Hence the sum of the mth powers is 0 unless we also have

$$1 - \omega_n^m = 0,$$

in which case the formula for the sum is indeterminate. But the only times $1 - \omega_n^m = 0$ is when m is a multiple of n:

$$1 - \omega_n^{nt} = 1 - (\omega_n^n)^t = 1 - 1 = 0.$$

For all other m, that is, for $m \equiv 1, 2, 3, \ldots, n - 1 \pmod{n}$, we have

$$1 + \omega_n^m + \omega_n^{2m} + \omega_n^{3m} + \cdots + \omega_n^{(n-1)m} = 0.$$

In the case of $m \equiv 0 \pmod{n}$, we have $m = nt$ for some integer $t \geq 0$, giving

$$(\omega_n^s)^m = \omega_n^{snt} = (\omega_n^n)^{st} = 1,$$

and

$$1 + \omega_n^m + \omega_n^{2m} + \omega_n^{3m} + \cdots + \omega_n^{(n-1)m}$$
$$= 1 + 1 + 1 + 1 + \cdots + 1$$
$$= n.$$

Thus the contribution to the sum $|\sum a_k^m|$ made by the terms associated with the nth roots of unity is

$$\left|\sum_{s=o}^{n-1}(r\omega_n^s)^m\right| = r^m(0) = 0$$

except when m is a multiple of n, in which case it is $r^m(n)$.

This makes it a simple matter to check out the 14-term sequence given by r times the square, cube, fourth, and fifth roots of unity. Since the sum of the mth powers of the fifth roots is zero for $m = 1, 2, 3, 4$, the only time the fifth roots contribute anything to the sums in question is when $m = 5$; similarly, the only contribution of the fourth roots is when $m = 4$; the only contribution of the cube roots is when $m = 3$, and none of the roots contribute anything for $m = 1$. The square roots, however, do make a contribution for each of $m = 2$ and $m = 4$. Thus we easily obtain, as above, that

$$\left|\sum a_k\right| = 0, \left|\sum a_k^2\right| = 2r^2,$$
$$\left|\sum a_k^3\right| = 3r^3, \left|\sum a_k^4\right| = 6r^4; \quad \text{and} \quad \left|\sum a_k^5\right| = 5r^5.$$

Thus, to finish, all we need to do is to choose r so that $6r^4 < 5r^5$. But this requires

$$\tfrac{6}{5} < r,$$

which comes as a terrible blow! Clearly we can't allow r even to be as great as 1, for the condition $|a_k| = r < 1$ is obligatory. For a while, though, things were looking pretty good.

But wait! The fifth roots don't come into play until $m = 5$, and so the only effect of including a second copy of the terms that are associated with the fifth roots is to double the value of $|\sum a_k^5|$ to $10r^5$. In this case, then, all we need is $6r^4 < 10r^5$, that is, $\tfrac{3}{5} < r$.

Not forgeting that $\tfrac{2}{3} < r$ is needed to make $2r^2 < 3r^3$, a successful 19-term sequence, containing an extra copy of the five terms associated with the fifth roots, is obtained for the five-link chain by any r in the range $\tfrac{2}{3} < r < 1$.

We observe that this chain of inequalities can similarly be extended to any number of links by taking as many copies as might be necessary of the terms associated with the various sets of nth roots.

The Solution to Problem 5

Clearly $\triangle PAB$ will be isosceles if $PA = PB$; however, it will also be isosceles if $PA = AB$ or $PB = AB$. Recalling that $\triangle ABC$ is equilateral, Figure 12 shows there are 10 positions for the point P.

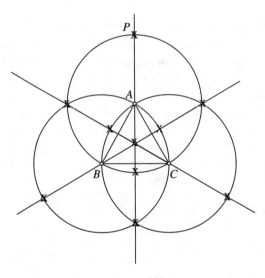

FIGURE 12

From *Leningrad Olympiads*

This section is based on selections from the book *Leningrad Mathematical Olympiads*, 1987–1991, by Dimitry Fomin and Alexey Kirichenko (MathPro Press, 1994, http://www.MathProPress.com), another outstanding collection. We touch on only a small sample of these intriguing problems.

1. A geometer wants to drop a perpendicular from a point P to a straight line m which does not pass through P (Figure 1). His only tools are

   (i) a straightedge,

   (ii) a tool that erects a perpendicular to a straight line **from a point on the line**.

   How can he perform the construction?

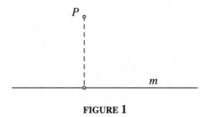

Solution 1. This solution is based on the following lemma, a neat result in its own right, which we shall prove at the end:

> *in* $\triangle ABC$, *if a point R on median AD is projected to Q and P on the opposite sides from B and C, the segment PQ is parallel to BC (Figure 2).*

Thus, if we can find the midpoint D of any segment BC along the line m, we can extend BP to any point A to obtain a triangle ABC whose median AD is known. Then, joining PC gives a point R on AD, which can be projected from

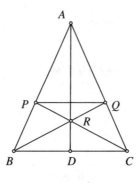

FIGURE 2

B to Q on AC, to give the segment PQ parallel to BC. Finally, the required line is obtained by drawing the perpendicular to PQ at P.

But finding the midpoint D of a segment BC on m is easy. First choose any point B on m and join BP with the straightedge (Figure 3). Then, with the special tool, draw a perpendicular to PB at P to meet m at C. Similarly, draw perpendiculars at B and C to PB and PC, respectively, to complete rectangle $PBEC$. Then diagonal PE bisects diagonal BC on m.

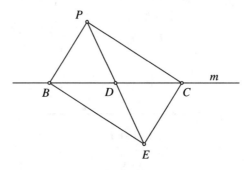

FIGURE 3

Proof of the Lemma: Let S be the midpoint of QC (Figure 4). Then, in $\triangle BQC$, DS joins the midpoints of two sides and is therefore parallel to the third side BQ. Hence RQ and DS are parallel in triangle ADS, giving

$$\frac{AQ}{QS} = \frac{AR}{RD}.$$

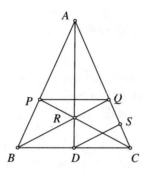

FIGURE 4

Since $QC = 2QS$, then

$$\frac{AQ}{QC} = \frac{AQ}{2QS} = \frac{AR}{2RD}.$$

Similarly, bisecting BP leads to

$$\frac{AP}{PB} = \frac{AR}{2RD},$$

and we have

$$\frac{AQ}{QC} = \frac{AP}{PB},$$

implying PQ and BC are parallel.

This solution is okay, but now consider the brilliant published solution.

As in solution 1, any line PQ parallel to m leads immediately to a solution. After drawing any rectangle $PBEC$ as in solution 1 (Figure 5), similarly draw

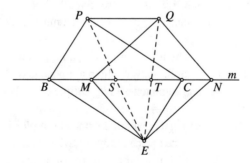

FIGURE 5

any other rectangle *ENQM* at *E*. Then *m* crosses diagonals *PE* and *QE* at their midpoints *S* and *T*, making *PQ* parallel to *ST*.

2. (A little teaser)

Is it possible to write the positive integers from 1 to 100 in a row so that adjacant numbers differ by at least 50?

(The answer is given at the end of this section.)

3. Two players, first *A* and then *B*, take turns removing matches from a pile that starts off with 500 matches. The player who takes the last match wins. The catch is that, on each turn, the number of matches that is withdrawn must be a power of 2. Is there a winning strategy for *A* or *B*?

It's a tantalizing puzzle to see how powers of 2 figure into the game, but the secret is revealed by taking the standard approach of considering games that start with a small number of matches. Let the number of matches at the beginning be *N*.

Clearly *A* wins on his first move if *N* is a power of $2(1, 2, 4, 8, \ldots)$.

$N = 3$, however, is a loser for *A*, for he must take either 1 or 2 matches, either of which leaves a power of 2 for *B*.

On the other hand, $N = 5$ wins for *A*, for he can remove 2 and leave *B* with a losing pile of 3 matches.

$N = 6$ is another loser for *A*: he must leave either 5, 4, or 2 matches for *B*, and we just saw that 5 is a winner for the one who receives it.

$N = 7$ wins for *A*, for it allows *A* to give *B* a pile of 6 matches which, as above, loses for the player who receives it.

$N = 9$, however, loses for *A*: he would have to leave *B* either 8, 7, 5, or 1 matches, all of which win for *B*.

Summarizing, a player wins if he receives 1, 2, 4, 5, 7, or 8 matches and loses with 3, 6, or 9 matches, suggesting that multiples of 3 are the key to the problem.

A positive integer which is itself not a multiple of 3 must exceed a multiple of 3 by either 1 or 2, both of which are powers of 2. Thus one can always reduce the number of matches to a multiple of 3 when it isn't one to begin with.

Conversely, subtracting a power of 2 from a multiple of 3 cannot result in another multiple of 3 since one can move between multiples of 3 only by subtracting an amount which is itself a multiple of 3, and that is something no power of 2 can manage.

Thus if A can once leave B with a pile in which the number of matches is a multiple of 3, B's move must take the number to a positive number which is not a multiple of 3, which A can immediately restore to a multiple of 3. Since B cannot move to 0 from a positive multiple of 3, he cannot win, and therefore A must win with this strategy.

Since 500 is not a multiple of 3, A is able to get things going by taking 2 on his first move, leaving B with $498 = 3 \cdot 166$ matches.

While the game hardly bears repetition, its analysis is a nice exercise.

4. Find all the integer solutions of the system of equations

$$ab + cd = -1 \tag{1}$$
$$ac + bd = -1 \tag{2}$$
$$ad + bc = -1. \tag{3}$$

From the symmetry of the system, it follows that, if (a, b, c, d) is a solution, so is each of the permutations (a, b, d, c), (a, c, b, d), ...; moreover, since each literal term in each equation contains two of the variables, $(-a, -b, -c, -d)$ is also a solution.

Now, if all four variables have the same value, the first equation would be $2a^2 = -1$, which has no integer solution. Thus, while not all variables can be the same, it is not difficult to see that some three of them are necessarily the same:

Subtracting (3) from (2) gives $(a - b)(c - d) = 0$, implying that $a = b$ or $c = d$. Similarly, (3) from (1) gives $(a - c)(b - d) = 0$, implying $a = c$ or $b = d$. Altogether, then, we have only the four possibilities

$$(a = b \text{ and } a = c), \qquad (a = b \text{ and } b = d),$$
$$(c = d \text{ and } a = c), \qquad (c = d \text{ and } b = d),$$

in each of which three of a, b, c, d are equal. Thus the solutions go together in families of eight:

$$(x, x, x, y), (x, x, y, x), (x, y, x, x), (y, x, x, x),$$

and their negatives.

In each family there is a solution of the form $(a, b, c, d) = (x, x, x, y)$, in which case equation (1) yields

$$x^2 + xy = -1,$$
$$x^2 + yx + 1 = 0,$$

giving

$$x = \frac{y \pm \sqrt{y^2 - 4}}{2}.$$

Since x and y are integers, $y^2 - 4$ must be a perfect square, making both y^2 and $y^2 - 4$ perfect squares. But the only two perfect squares that differ by 4 are 0 and 4, and therefore

$$y^2 = 4, \quad y = \pm 2, \quad \text{giving} \quad x = \frac{y}{2} = \mp 1.$$

Thus there is only one family of solutions, namely the four permutations of $(-1, -1, -1, 2)$ and their negatives.

5. Is it possible to arrange a set S of 100 consecutive positive integers around a circle in such a way that the product of each pair of adjacent numbers is a perfect square?

While one can hardly believe such a scheme is possible, it is not obvious how to formulate a proof. If each integer is written in the form $2^r q$, where q is odd, an odd integer would have $r = 0$ and an even integer would have $r \geq 1$. Since the even numbers proceed $2, 4, 6, \ldots, 4k, 4k + 2, \ldots$, every second even number is of the form $4k + 2 = 2(2k + 1) = 2^1 q$, where q is odd. Thus, while the fifty odd numbers in S have $r = 0$, half its fifty even numbers have an *odd* exponent $r = 1$.

Now, in order for the product of adjacent numbers to be a perfect square, a number with an odd exponent r must be coupled with another like itself. Thus all the numbers having odd r must occur together around the circle in an unbroken string. But, not comprising all 100 numbers in S, this string cannot extend all the way around the circle. The last member of the string and its immediate outside neighbor have a product of the form

$$2^{\text{odd}} q \cdot 2^{\text{even}} k,$$

containing an unacceptable odd power of 2.

6. Given 32 stones, each a different weight, show how to determine both the heaviest stone and the second heaviest stone in 35 weighings with an equal-arm balance.

Suppose we liken the stones to tennis players, the heavier the stone the better the player, and that they are the 32 players in a singles tournament. Suppose also that there are no upsets; the better player always wins. Each match corresponds to a use of the equal-arm balance—the better player goes on to the next round in the tournament, corresponding to the heavier stone going through to another round of weighings.

With 32 players, the tournament would have five rounds consisting successively of 16, 8, 4, 2, and 1 matches, for a total of 31 matches (each match produces a loser and 31 of the 32 players have to be eliminated).

Now, nobody is good enough to beat the best player and so, no matter who plays who in the various rounds, he will be the eventual winner. Thus the best player (heaviest stone) is easily determined in 31 matches (weighings). The interesting problem is finding the second best player.

Since the second best player didn't win the tournament, he must have been eliminated sometime. But the only player good enough to beat him is the best player himself, and so these two must have squared off in some match of the tournament. By keeping a record of the pairings throughout the tournament, we can identify the five players who were eliminated by the champion, and one of these is the second best player.

Now, if these five players were to have a little tournament among themselves, the second best player would win it, and the other four would lose. Thus another four matches suffice to identify the second best player, bringing the total to 35 matches.

7. Now for a remarkable result.

> Is there a set of 100 different positive integers such that, for any selection of five of them, their product is divisible by their sum?

We follow the published solution.

Let $S = \{a_1, a_2, \ldots, a_{100}\}$ be any 100 different positive integers. Determine the $\binom{100}{5}$ sums of these numbers taken 5 at a time and let X be the product of these more than 75 million sums. Then X is divisible by each of the sums $(a_i + a_j + a_k + a_m + a_n)$,

$$(a_i + a_j + a_k + a_m + a_n)|X,$$

and X times a sum divides X^2:

$$(a_i + a_j + a_k + a_m + a_n)X|X^2.$$

Thus the sum of the five integers

$$(a_i X), (a_j X), (a_k X), (a_m X), (a_n X),$$

namely the number

$$(a_i + a_j + a_k + a_m + a_n)X,$$

being a divisor of X^2, clearly divides their product

$$a_i a_j a_k a_m a_n X^5,$$

and it follows that the 100 numbers $\{a_1 X, a_2 X, \ldots, a_{100} X\}$ possess the desired property.

8. The perimeter of the star polygon $PQRST$ in Figure 6 is 1:

$$PQ + QR + RS + ST + TP = 1.$$

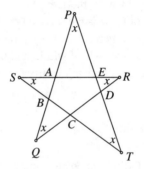

FIGURE 6

If the five angles x at the vertices P, Q, R, S, T are equal, determine the perimeter p of the convex pentagon $ABCDE$.

Clearly, the equal angles x at Q and R make $\triangle AQR$ isosceles; hence exterior angle $SAB = 2x$. Similarly, $\angle SBA = \angle P + \angle T = 2x$, and $\triangle SAB$ is isosceles. Thus the angles of $\triangle SAB$ are x, $2x$, and $2x$, and since they add up to $180°$, they are 36, 72, and 72 degrees.

Now, it is well known that the ratio of an arm to the base in such a triangle is the famous golden ratio

$$\tau = \frac{1 + \sqrt{5}}{2}$$

(the easy proof is given at the end). Hence

$$SA = SB = \tau AB, \quad \text{and} \quad SA + SB = 2\tau AB.$$

Similarly around the figure, $QB + QC = 2\tau BC$, and so forth. The left sides of these five equations plus the perimeter p of the pentagon add up to the entire perimeter of the polygon. Hence

$$1 = 2\tau(AB + BC + CD + DE + EA) + p$$
$$= 2\tau p + p = p(2\tau + 1),$$

and so

$$p = \frac{1}{2\tau + 1} = \frac{1}{2 + \sqrt{5}} \cdot \frac{\sqrt{5} - 2}{\sqrt{5} - 2} = \sqrt{5} - 2.$$

Proof that $SA = \tau AB$: In Figure 7, let BF bisect the $72°$ angle at B. Then $\angle FBA = 36°$, and since $A = 72°$, then $\angle BFA$ is also $72°$, and triangles SAB and ABF are similar. Thus

$$\frac{SA}{AB} = \frac{AB}{AF}. \tag{1}$$

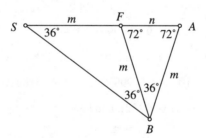

FIGURE 7

Let $SF = m$ and $FA = n$. Now, triangles SBF and ABF are both isosceles, and so $SF = FB = AB = m$. Consequently, $SA = m + n$, and (1) gives

$$\frac{m + n}{m} = \frac{m}{n},$$
$$m^2 - mn - n^2 = 0,$$
$$\left(\frac{m}{n}\right)^2 - \frac{m}{n} - 1 = 0,$$

and

$$\frac{SA}{AB} = \frac{m}{n} = \frac{1 + \sqrt{5}}{2} = \tau,$$

the positive root.

9. (A mind-boggler; factorial ten million is a BIG number!)

Clearly there are 10 million 7-digit nonnegative integers when the ones with fewer digits are padded on the left with 0's:

$$0000000, 0000001, 0000002, \ldots, 9999999.$$

Suppose these are concatenated in any order to form a 70-million-digit number N. Prove that each of these $10000000!$ gigantic numbers N is divisible by 239.

We follow the published solution.

If this amazing proposition is true, then the difference between any two of these numbers must also be divisible by 239. In particular, consider an N whose last two blocks are (0000001) and (0000000), and the number N' obtained from N by interchanging the final two blocks:

$$N' = \cdots (0000000)(0000001)$$
$$N = \cdots (0000001)(0000000).$$

Since their difference is

$$9999999 = 10^7 - 1,$$

we infer that 239 must also divide $10^7 - 1$. Checking, we find that

$$10^7 - 1 = 9999999 = 239 \cdot 41841.$$

It follows that $10^7 \equiv 1 \pmod{239}$, and hence also all powers

$$(10^7)^n \equiv 1 \pmod{239}.$$

Now, N is the sum of the "0-padded" 7-digit numbers $0, 1, 2, 3, \ldots,$ 9999999, where each is multiplied by the power of 10^7 that is called for by its position in the concatenation:

$$N = 0000000 \cdot (10^7)^a + 0000001 \cdot (10^7)^b + 0000002 \cdot (10^7)^c$$
$$+ \cdots + 9999999(10^7)^k.$$

Recalling that $\pmod{239}$, $(10^7)^n \equiv 1$ for all exponents n, then

$$N \equiv 0 \cdot 1 + 1 \cdot 1 + 2 \cdot 1 + \cdots + 9999999 \cdot 1 \pmod{239}$$
$$= 1 + 2 + 3 + \cdots + 9999999$$
$$= \frac{(9999999)(10000000)}{2}$$
$$= 239 \cdot 41841 \cdot 5000000$$
$$\equiv 0 \pmod{239},$$

establishing this remarkable result. In fact, N is always divisible by 9999999 and hence also by each of its divisors.

10. It is known that there is one counterfeit coin X among a certain set of 101 coins. The 100 genuine coins all have the same weight, which is different from the weight of X. How can it be determined whether X is heavier or lighter in two weighings with an equal-arm balance?

    Let the coins be divided arbitrarily into three subsets A, B, and C containing, respectively, 30, 30, and 41 coins, and then weigh A against B.

    (i) If A and B balance, they must contain only genuine coins since there is only one counterfeit coin, and X must be in C. Hence the behavior of C, when weighed against 41 genuine coins from the 60 in A and B, reveals whether X is heavier (if C goes down) or lighter (if C goes up).

    (ii) Suppose a balance is not obtained. For definiteness, suppose A goes down. Then C contains only genuine coins, and either X is in A and is heavier, or X is in B and is lighter. Thus, finding **where** X is also reveals what **kind** of coin it is.

    Clearly if X is not in A, it must be in B. To find out whether X is in A, simply divide A in half and weigh the two halves against each other—if they balance, then A contains only genuine coins and X must be in B; if they don't, then X is in A.

    Comment: The 30-30-41 split into the subsets A, B, and C is not critical. The size of the equal subsets A and B can be any even number from 26 to 50. It is only necessary that A and B together contain as many coins as C, in order to decide the issue when A and B balance, and that A have an even number of coins so it can be divided in half to settle the matter when A and B don't balance:

    $$|A| + |B| \geq |C| \quad \text{and} \quad |A| = |B| = \text{an even nunber.}$$

11. Prove that, in every arithmetic progression of positive integers,

    $$a, a + d, a + 2d, \ldots,$$

    there are two terms whose digits have the same sum.

    If the number of digits in a is k, then $10^k d + a$ is the number formed by concatenating the digits of d and a:

    $$10^k d + a = \underset{d}{(\ldots)} \underset{a}{(\ldots)}$$

Similarly,

$$10^{k+r}d + a = (\ldots) \underbrace{(000\ldots00)}_{r\,0\text{'s}} (\ldots)$$

where there are r 0's between the d and the a. Hence all the terms of the progression of the form $a + 10^{k+r}d$ have the same sum, namely the sum of the digits of a and d.

12. Let A be the set of all subsets of $N = \{1, 2, 3, \ldots, n\}$ which do not contain two consecutive integers. For example, for $\{1, 2, 3, 4\}$, we would have

$$A = \{\{1\}, \{2\}, \{3\}, \{4\}, \{1, 3\}, \{1, 4\}, \{2, 4\}\}.$$

Now multiply the members of each subset in A to obtain a collection of products P:

$$P = \{1, 2, 3, 4, 3, 4, 8\}.$$

Next, square each member of P to get a collection of squares S:

$$S = \{1, 4, 9, 16, 9, 16, 64\}.$$

Finally, add up all the squares in S to get a total T:

$$T = 1 + 4 + 9 + 16 + 9 + 16 + 64 = 119.$$

For $n = 4$, then, we have $T = 119$, which we observe is $5! - 1$.

Prove the remarkable fact that T is always $(n + 1)! - 1$.

This result is certainly remarkable, but its proof is an easy application of induction.

The case of $n = 1$ is trivial:

$$A = P = S = \{1\} \quad \text{and} \quad T = 1 = 2! - 1.$$

Suppose the claim is valid for $n = 1, 2, \ldots, k - 1$, and consider the case of $N = \{1, 2, 3, \ldots, k\}$.

A subset in A either contains the integer k or it doesn't. Those which do not contain k are the pertinent subsets of $\{1, 2, \ldots, k - 1\}$, for which the induction hypothesis gives a total of $T_1 = k! - 1$.

Now, since consecutive integers are forbidden, a subset of A which contains k is either just the integer k itself or it consists of k in combination with an acceptable nonempty subset of $\{1, 2, \ldots, k - 2\}$. For $\{1, 2, \ldots, k - 2\}$, the induction hypothesis gives a total of $(k - 1)! - 1$, and if k is attached to each of its acceptable subsets, the multiplying, squaring, and adding would result in a total that would be k^2 times as much. That is to say, the subtotal that is due to the subsets which contain k in combination with a nonempty subset of $\{1, 2, \ldots, k - 2\}$ is

$$k^2 \cdot [(k - 1)! - 1] = k \cdot k! - k^2.$$

Not forgetting the k^2 which is contributed by the subset $\{k\}$ containing just the integer k, we have altogether that the subsets containing k give a subtotal of

$$T_2 = (k \cdot k! - k^2) + k^2 = k \cdot k!.$$

The grand total T for $\{1, 2, \ldots, k\}$ is therefore

$$\begin{aligned}
T &= T_1 + T_2 \\
&= (k! - 1) + k \cdot k! \\
&= (k + 1) \cdot k! - 1 \\
&= (k + 1)! - 1,
\end{aligned}$$

and the conclusion follows by induction.

A Solution to #2:

$$51 \quad 1 \quad 52 \quad 2 \quad 53 \quad 3 \quad \ldots \quad 48 \quad 99 \quad 49 \quad 100 \quad 50.$$

From *The Contest Problem Book V*

The contests which are covered in this excellent book (prepared by George Berzsenyi and Stephen Maurer, The Anneli Lax New Mathematical Library Series, volume 38, 1997) are the American High School Mathematics Examinations (AHSME) and the American Invitational Mathematics Examination (AIME) for the years 1983–1988. A great many interesting problems and clever solutions are presented and the exposition is so well done that the book is a genuine pleasure to read.

The problems on these examinations are posed in multiple-choice form. We shall not retain this feature but simply enjoy them as straightforward challenges.

1 From AHSME Contests

1. (1983, #22)

Consider the two functions

$$f(x) = x^2 + 2bx + 1 \quad \text{and} \quad g(x) = 2a(x + b),$$

where the variable x and the constants a and b are real numbers. Each pair of constants (a, b) may be considered to be the coordinates of a point in an ab-plane. Let S be the set of all such points (a, b) for which the graphs of $y = f(x)$ and $y = g(x)$ do **not** intersect.

Determine the area of S.

Algebraically the requirement that the graphs do not intersect is that the equation $f(x) = g(x)$ should have no real root; that is to say, the discriminant of

$$x^2 + 2bx + 1 = 2a(x + b)$$

that is, of

$$x^2 + (2b - 2a)x(1 - 2ab) = 0,$$

should be negative. This would confine (a, b) to the region S given by

$$(2b - 2a)^2 - 4(1 - 2ab) < 0,$$
$$(b - a)^2 - 1 + 2ab < 0,$$
$$a^2 + b^2 < 1,$$

which is just the interior of the unit circle centered at the origin. Hence the area of S is π.

2. (1984, #28)

How many distinct pairs of positive integers (x, y) are there such that

$$x < y \quad \text{and} \quad \sqrt{1984} = \sqrt{x} + \sqrt{y}?$$

The prime decomposition of 1984 is $2^6 \cdot 31$, making

$$\sqrt{1984} = 8\sqrt{31}.$$

Now,

$$\sqrt{y} = \sqrt{1984} - \sqrt{x},$$

and squaring gives

$$y = 1984 + x - 2\sqrt{1984x}$$
$$= 1984 + x - 16\sqrt{31x}.$$

Since y is an integer, x must be of the form $31t^2$.

But, since $x < y$, then $\sqrt{x} < \sqrt{y}$, and therefore $\sqrt{x}$ is less than half the sum $\sqrt{x} + \sqrt{y} = 8\sqrt{31}$. Thus $\sqrt{x} < 4\sqrt{31}$ and $x < 31 \cdot 4^2$, leaving x only the three values $31 \cdot 1^2$, $31 \cdot 2^2$, $31 \cdot 3^2$. Hence there are only three pairs (x, y).

3. (1987, #16)

A cryptographer devises the following method of encoding positive integers. First, a 1-1 correspondence is set up between the digits 0, 1, 2, 3, 4 and the letters in the set $\{V, W, X, Y, Z\}$, for example,

$$
\begin{array}{ccccc}
0 & 1 & 2 & 3 & 4 \\
\downarrow & \downarrow & \downarrow & \downarrow & \downarrow \\
Y & V & X & Z & W.
\end{array}
$$

Second, the integer to be encoded is expressed in base 5.

With the particular correspondence chosen by the cryptographer, there were three consecutive integers in increasing order that were encoded, respectively, by VYZ, VYX, and VVW.

What is the base ten number that would thus be encoded XYZ?

Since VYZ, VYX, and VVW represent consecutive integers, the sums

$$
\text{(i)} \quad
\begin{array}{ccc}
V & Y & Z \\
  &   & 1 \\
\hline
V & Y & X
\end{array}
\quad \text{and} \quad
\text{(ii)} \quad
\begin{array}{ccc}
V & Y & X \\
  &   & 1 \\
\hline
V & V & W
\end{array}
$$

are valid in base 5.

Because the "5's" digit, Y, remains unchanged in (i), it follows that there can't be a carryover to the second place and $X = Z + 1$. However, in (ii), the 5's digit *does* change. Thus, in (ii), there must be a carryover to the second place, implying $W = 0$ and $X = 4$. This makes $Z = X - 1 = 3$. It remains to assign the digits 1 and 2 to Y and V. In view of the carryover in (ii), $Y = 2$ would make $V = 3$, not 1. Hence it must be that $Y = 1$ and $V = 2$. The selected correspondence, then, is

$$
\begin{array}{ccccc}
0 & 1 & 2 & 3 & 4 \\
\downarrow & \downarrow & \downarrow & \downarrow & \downarrow \\
W & Y & V & Z & X.
\end{array}
$$

The code XYZ thus represents the integer 413 in base 5, which is 108 in base 10.

2 From AIME Contests

4. (1984, #11)

A gardener plants three maple trees, four oak trees, and five birch trees in a row. He plants them in random order, with each arrangement being equally likely.

What is the probability that no two birch trees are next to one another?

(i) Let's assume that no two trees are identical, and so the planting involves 12 different trees, for a total of 12! possible arrangements.

Now, there are 7! ways of arranging the 7 different maple and oak trees in the row. In order to avoid adjacent birch trees, the birch trees must be placed one in each of 5 of the 8 places around the maple and oak trees:

$$\downarrow \quad \downarrow \quad \downarrow \quad \downarrow \quad \downarrow \quad \downarrow \quad \downarrow \quad \downarrow$$
$$m \quad m \quad o \quad o \quad o \quad m \quad o \quad .$$

This can be done in $8 \cdot 7 \cdot 6 \cdot 5 \cdot 4$ ways, and the required probability is

$$\frac{7! \cdot 8 \cdot 7 \cdot 6 \cdot 5 \cdot 4}{12!} = \frac{7 \cdot 6 \cdot 5 \cdot 4}{12 \cdot 11 \cdot 10 \cdot 9} = \frac{7}{99}.$$

(ii) Now, one might feel uneasy about having assumed the trees to be all different, and might feel better if the problem were done again on the basis that, while any maple tree is certainly different from any oak tree or any birch tree, it was three identical maple trees that were planted among four identical oak trees and five identical birch trees.

In this case, there are not 7! ways of planting the 7 maple and oak trees, but only

$$\frac{7!}{3!4!}$$

ways. The number of ways of putting the birch trees in the spaces between them would only be $\binom{8}{5}$. Noting that the total number of possible arrangements in this case is only

$$\frac{12!}{3! \cdot 4! \cdot 5!},$$

and that $\binom{8}{5} = 56 = 8 \cdot 7$, we obtain the probability of no consecutive birch trees to be

$$\frac{\frac{7!}{3!\cdot4!} \cdot \binom{8}{5}}{\frac{12!}{3!\cdot4!\cdot5!}} = \frac{7! \cdot 8 \cdot 7 \cdot 5!}{12!} = \frac{7 \cdot 5 \cdot 4 \cdot 3 \cdot 2}{12 \cdot 11 \cdot 10 \cdot 9} = \frac{7}{99},$$

the same as before. (Remarkable!)

From *Quantum*

1. First a lovely Problem in Combinatorics (A problem from V. Proizvolov's excellent article "Problems Teach Us How To Think," Jan./Feb., 2002, page 42–45.)

> In a certain chess position, the number of pieces in each horizontal row and the number of pieces in each vertical column is an odd number. Prove that the total number of pieces on black squares is an even number.

As usual in chess, let the bottom left corner be black. I'm sure I could look at a chessboard for a long time without it dawning on me that, numbering the rows from the bottom and the columns from the left, every black square is either in an odd-numbered row or in an even-numbered column, that is, that all the black squares are corralled in rows 1, 3, 5, 7, and columns 2, 4, 6, 8.

In celebration of this triumphant observation, let the black squares in rows 1, 3, 5, 7, each be marked with the letter A and each black square in columns 2, 4, 6, 8 be marked with the letter B. Also, let the white squares in rows 1, 3, 5, 7 be marked with the letter C, leaving the rest of the white squares unmarked (Figure 1). Finally, let a, b, c, denote the total number of

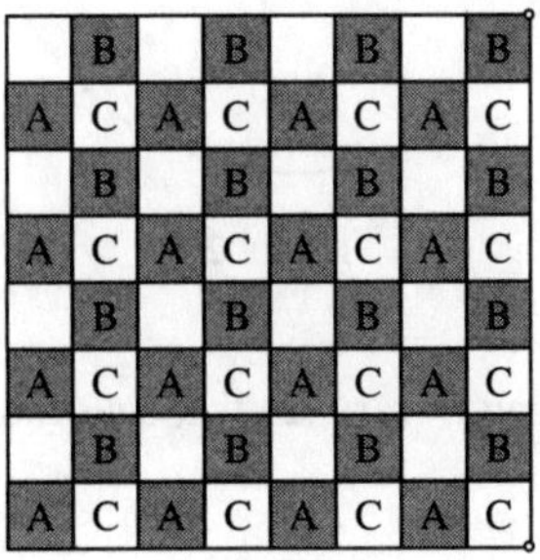

FIGURE 1

pieces which sit on squares marked A, B, C, respectively. We want to show that $a + b$ is an even number.

Now, each row contains an odd number of pieces. Thus $a + c$, which is the total number of pieces in rows 1, 3, 5, 7, is the sum of four odd numbers, and is therefore an even number. Similarly, $b + c$, which is the total number of pieces in columns 2, 4, 6, 8, is also the sum of four odd numbers and hence an even number. Therefore

$$(a + c) + (b + c) = a + b + 2c$$

is an even number, implying the desired $a + b$ is even.

2. A Problem of E. Kulanin (Problem M312, Jan./Feb., 2001, page 11)

MN is a fixed chord in a given circle S with center O, and AB is a variable diameter that does not intersect MN. AM and BN meet at C outside the circle (Figure 2). Thus, for each position of AB, there is a triangle ABC and an altitude CD to AB.

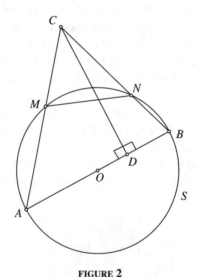

FIGURE 2

Prove that, as AB and C vary, all these altitudes are concurrent.

Since AB is a diameter, angles AMB and ANB are right angles, and so AN and BM are altitudes in $\triangle ABC$. Thus AN and BM intersect at the orthocenter H of $\triangle ABC$, and the third altitude CD also goes through H (Figure 3).

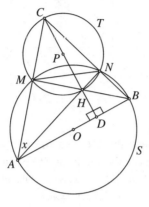

FIGURE 3

Now, MN subtends the same angle at each point A on the circumference of S, and therefore, in each triangle CAN, the angles at A and N never change. Hence $\angle C$ must also be constant, implying that MN always subtends the same angle at C. Thus C lies on a fixed circle T through M and N.

Because of the supplementary right angles at M and N, the circle T through C, M, and N is simply the circle on diameter CH. Therefore, as AB rotates about its center O, C and H move around T at the ends of a diameter. Thus CH always goes through the center P of T, and since the altitude CD lies along CH, all the altitudes CD are concurrent at P.

3. The 1996 May–June issue of *Quantum* featured an article by Vladimir Dubrovsky (University of Moscow) in which he presented a parade of ingenious approaches to a problem of V. Shafaryan that appeared in an old issue of Kvant, the parent of *Quantum*. The article is called "Cutting Facets." Here is the problem.

    A rectangle $KLMN$ is inscribed in a circle (Figure 4). From an arbitrary point P on the circle, straight lines are drawn parallel to the sides of the rectangle to meet one pair of opposite sides of the rectangle at A and B and the extensions of the other pair of sides at C and D.

    Prove that AC and BD are perpendicular and that they meet on the diagonal KM.

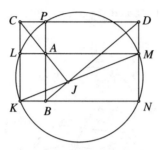

FIGURE 4

(a) (i) First let us show that AC and BD are perpendicular. Let PB be extended to meet the circle at Q (Figure 5). Then, by symmetry, $BQ = PA = CL = DM(= s)$ and $KB = CP(= t)$. Also $AB = MN(= u)$, and let $BN = v$.

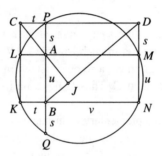

FIGURE 5

Then, relative to coordinate axes in the directions of the sides of rectangle $KLMN$,

$$\text{the slope of } BD = \frac{s+u}{v},$$

and

$$\text{the slope of } AC = -\frac{s}{t}.$$

Hence we would like to show that

$$\left(\frac{s+u}{v}\right)\left(-\frac{s}{t}\right) = -1, \text{ i.e., that } s(s+u) = vt.$$

But this is immediate from chords PQ and KN which intersect at B (Figure 6).

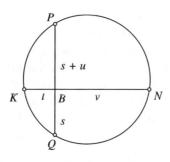

FIGURE 6

(ii) Now for a delightful surprise. The perpendicularity of *AC* and *BD* follows almost immediately from the famous theorem of Brahmagupta:

> *If a cyclic quadrilateral has perpendicular diagonals, then a line from the midpoint of a side, through the point of intersection of the diagonals, is perpendicular to the opposite side.*
> (This is proved in the Appendix to this section.)

Clearly *PLQM* is a cyclic quadrilateral with perpendicular diagonals *PQ* and *LM* (Figure 7). Since the diagonals of rectangle *CLAP* bisect each other, the line *AC* does in fact go through the midpoint *T* of *PL*, and it follows by Brahmagupta's theorem that *CA* is perpendicular to *QM*. But, since *BQ* and *MD* are equal and parallel, *BQMD* is a parallelogram and any line perpendicular to *QM* is also perpendicular to *BD*.

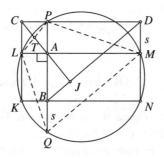

FIGURE 7

(b) We conclude with the easy proof that AC and BD meet on KM.

Let AC meet BD at J (Figure 8). We have just proved that CA and BD are perpendicular, and so $\angle CJB$ is a right angle. Hence BC subtends a right angle at both P and J, implying that the circle on diameter BC goes through J. But clearly, this circle is the circumcircle of rectangle $CKBP$, and therefore KP is also a diameter of the circle. Accordingly, KP also subtends a right angle at J, i.e., $\angle PJK = 90°$.

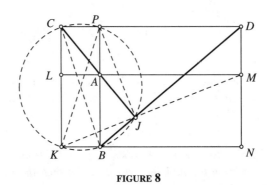

FIGURE 8

Similarly, AD and PM are diameters of the circumcircle of $PAMD$. We know that AD subtends a right angle at J, implying the circumcircle goes through J. Thus diameter PM also subtends a right angle at J, making $\angle PJM = 90°$.

Being comprised of two right angles, $\angle KJM$ is a straight angle, and the conclusion follows.

4. Our next problem comes from the excellent article "Surprises in Conversion" by I. Kushnir, Mar–Apr, 1996.

Recall that the Euler line of a triangle is the straight line that contains the orthocenter and the circumcenter.

It is easy to see that the Euler line of an isosceles triangle passes through the incenter of the triangle.

Proof: Suppose $AB = AC$. Then the altitude AD is actually the perpendicular bisector of BC (Figure 9). Since circumradii OB and OC are equal, then altitude AD not only contains the orthocenter H but also the circumcenter O, making it the Euler line of the triangle.

Moreover, AD is the bisector of angle A, and so it also passes through the incenter I, and the conclusion follows.

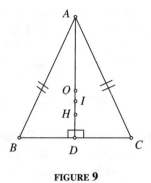

FIGURE 9

Prove the converse theorem:

If the incenter of a triangle lies on its Euler line, the triangle is isosceles.

As we saw in the direct theorem, the Euler line can be an angle bisector of a triangle. However, at least two of the angle bisectors must be distinct from the Euler line, say the bisectors of angles A and B. Suppose the bisectors AI and BI of angles A and B meet the circumcircle at D and E (Figure 10), and let the altitudes from A and B be AU and BT. Since AD bisects $\angle A$, then arcs CD and DB are equal. Similarly, arc $CE =$ arc EA and hence D and E are the midpoints of the arcs CB and CA. The radii DO and EO, then, are the perpendicular bisectors of the sides CB and CA, making them respectively parallel to the altitudes AU and BT. With OD parallel to HA, triangles ODI and IHA are similar, as also are triangles OEI and IHB.

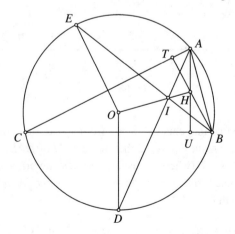

FIGURE 10

With sides *OI* and *HI* in each of these pairs of triangles, the ratio of corresponding sides is *OI/IH* for each pair. Thus *HA* is the same fraction of *OD* that *HB* is of *OE*. Since radii *OD* and *OE* are equal, then so must *HA* = *HB*. Thus triangle *HAB* is isosceles and

$$\angle HAB = \angle HBA.$$

But, in right triangle *AUB*, $\angle HAB = 90° - \angle B$, and in right triangle *BAT*, $\angle HBA = 90° - \angle A$. Hence $90° - \angle B = 90° - \angle A$, giving $\angle B = \angle A$ and making triangle *ABC* isosceles, as desired.

5. (Math Investigations: Geometry in the Pagoda, Jan.–Feb., 1995)

Inside △*ABC*, a small circle is tangent to sides *AB* and *AC* and its tangents *BD* and *CE* cross at *F*, as in Figure 11. Prove that the incircle of triangles *ABC* and *FBC* touch *BC* at the same point *X*.

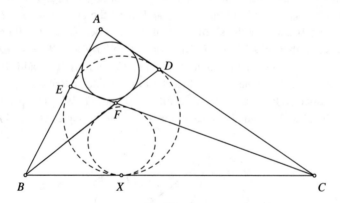

FIGURE 11

Since the two tangents to a circle from an external point are the same length, it is clear from Figure 12 that the incircle of △*ABC* meets *BC* at *X* such that $BX = s - b$ (where, as usual, *s* is the semiperimeter and $AC = b$).

Thus we need to show that "$s - b$" is the same for △*ABC* as it is for △*FBC*, that is, that

$$\frac{AB + BC + AC}{2} - AC = \frac{FB + BC + FC}{2} - FC,$$

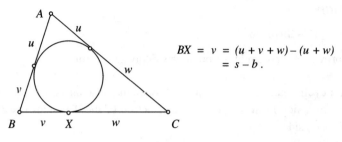

$$BX = v = (u + v + w) - (u + w)$$
$$= s - b.$$

FIGURE 12

which easily reduces to

$$AB - AC = FB - FC.$$

But this is easy.

Let the given circle touch its tangents at P, Q, R, S (Figure 13).

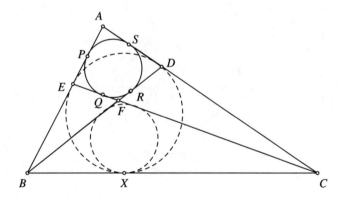

FIGURE 13

Then

$$
\begin{aligned}
AB - AC &= (AP + PB) - (AS + SC) \\
&= BP - CS \quad \text{(since } AP = AS\text{)} \\
&= BR - CQ \quad \text{(since } BP = BR \text{ and } CS = CQ\text{)} \\
&= (FB + FR) - (FC + FQ) \\
&= FB - FC \quad \text{(since } FR = FQ\text{), as desired.}
\end{aligned}
$$

Appendix

Bramagupta's theorem

Let us prove the theorem in the obviously equivalent form

> In a cyclic quadrilateral having perpendicular diagonals, the perpendicular to a side from the point of intersection of the diagonals bisects the opposite side.

Proof: Let TE be perpendicular to AD and let the acute angles in right triangle AET be x and y (Figure 14). Since the diagonals are perpendicular, $\angle DTE = x$, and in right triangle DTE the angle at D is y. The pairs of vertically opposite angles at T then give the further angles x and y shown there. But the chord CD subtends equal angles x at A and B on the circumference, and chord AB similarly gives equal angles y at C and D. Thus the triangles TQC and TQB are both isosceles, and the arms QC and QB are equal, since each is equal to the common arm QT.

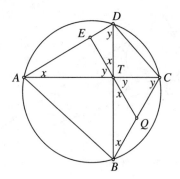

FIGURE 14

From *The Mathematical Visitor*

The Mathematical Visitor is an American journal that was published during the years 1877 to 1896. Stanley Rabinowitz has done the mathematical community a great service by publishing the collection *Problems and Solutions From The Mathematical Visitor* (MathPro Press, 1996, http://www.MathProPress.com).

1. (Problem 93, by E.J. Edmunds, 1879)

   P is an arbitrary point on the side BC of $\triangle ABC$. Determine how to draw a segment QR across the triangle which is parallel to BC and which subtends a right angle at P (Figure 1).

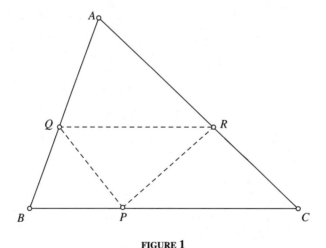

FIGURE 1

Obviously the circle on diameter QR goes through P. It is equally obvious that it is hopeless to try to guess the position of Q on AB. However, as it turns out, any guess is as good as the real thing.

From any point S on AB draw ST parallel to BC and let the semicircle on ST cross AP at L (Figure 2). Then QR is found simply by drawing PQ and PR respectively parallel to LS and LT.

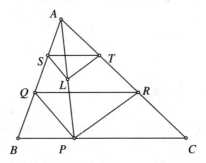

FIGURE 2

Clearly this makes $\angle QPR = \angle SLT = 90°$, and the parallel lines yield

$$\frac{AS}{SQ} = \frac{AL}{LP} = \frac{AT}{TR},$$

implying QR is parallel to ST, which was constructed parallel to BC.

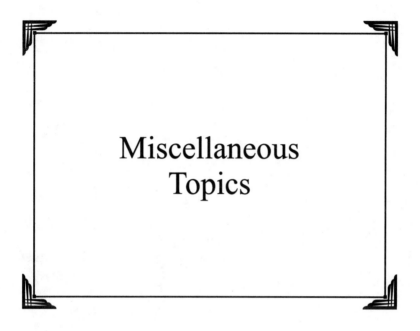

Miscellaneous
Topics

From the Desk of Liong-shin Hahn

From time to time a reader will send me an alternative solution to a problem in one of my essays, often a solution which is markedly superior to the published solution. It is a pleasure to present some inspired work by Professor Liong-shin Hahn, University of New Mexico, retired. Dr. Hahn is the author of the excellent volume *Complex Numbers and Geometry* (MAA, Spectrum Series, 1994).

1. A Safe-Cracking Problem

This intriguing problem comes from the final round of the 2002 New Mexico Mathematics Contest for high school students.

You are given a safe with a lock consisting of a 4 by 4 arrangement of keys. Each of the 16 keys can be in a horizontal or a vertical position. To open the safe all the keys must be in the vertical position. When you turn a key, all the keys in the same row and column change positions. You may turn a key more than once.

(a) Given the configuration in Figure 1, what is the minimum number of keys that must be turned to open the safe? Which keys must be turned?

(b) If you are allowed to turn at most 2002 keys, what is the largest $2n \times 2n$ safe that you can always open?

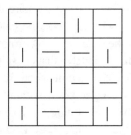

FIGURE 1

I hope you will have some fun with this problem before going on to the solution.

(a) Since turning a key also reverses the position of the 6 other keys in its row and column, trying to maintain a mental image of the overall state of affairs after turning a few keys is enough to discourage all but the stoutest heart. Still, with patience, it is not beyond trial and error to uncover the solution to part (a) (that is to say, that's the only way I could do it). Labelling the keys $a, b, \ldots, p$, as in Figure 2, the solution is to turn keys b and k. Since it is easy to check that no single key opens the safe, the solution bk is minimal.

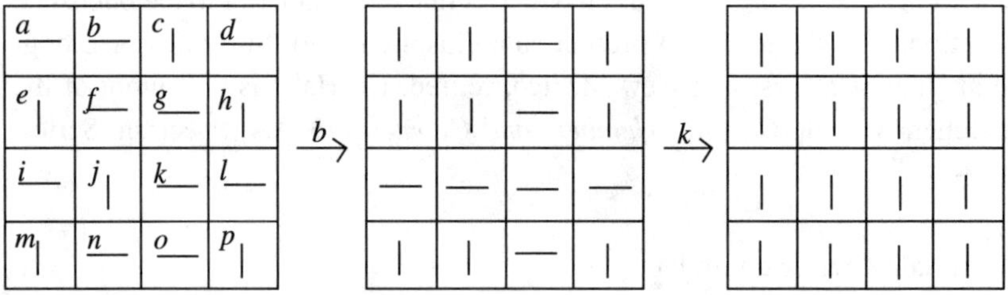

FIGURE 2

However, this approach is no good in general. What we need is a minimal safe-cracking procedure that always works.

(b) Clearly, turning a key twice restores each of the keys that it affects to its original position. Therefore turning a key an even number of times is equivalent to not turning it at all, and turning a key an odd number of times has the same effect as turning it once. Also, it is evident that, as a result of turning a succession of keys, the total number of times the state of a given key is reversed is the same regardless of the order in which the keys might be turned.

I don't expect it has taken you very long to come to the realization that, for an m by m safe, where m is *even*, the state of a key can be reversed, without changing any of the other keys, by turning each key in its row and column. This is not true for m odd, but that doesn't matter since our problem specifically concerns safes of even dimension. To see this in the 4 by 4 case, suppose we would like to reverse key a without altering any of the others (Figure 3). This is achieved by turning the 7 keys a, b, c, d, e, i, m. This changes key a 7 times, thus leaving it reversed at the end, but each other key is reversed the even number of times that is indicated in Figure 4.

a	b	c	d
e	f	g	h
i	j	k	l
m	n	o	p

7	4	4	4
4	2	2	2
4	2	2	2
4	2	2	2

FIGURE 3   FIGURE 4

Therefore, applying this 7-step procedure to each of the 10 horizontal keys in part (a), we obtain a 70-key string that opens the safe. Without writing out the entire string, we can see it contains 4 a's, 5 b's, 6 c's, and so on, as indicated in Figure 5: for example, key a gets turned in the process of reversing each of the 4 horizontal keys a, b, d, i, and only then; b only gets turned in reversing each of the 5 horizontal keys a, b, d, f, n, and so on.

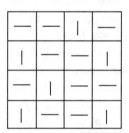

a	b	c	d
e	f	g	h
i	j	k	l
m	n	o	p

4	5	6	4
4	4	4	4
4	6	5	4
4	4	4	4

FIGURE 5

Thus the 60 turnings of keys which are turned an even number of times, namely the keys a, c, d, e, f, g, h, i, j, l, m, n, o, p, play no role in opening the safe. It's only turning b and k that has any effect, and since turning each of them once is as good as 5 times, our 70-key string boils down to just bk.

Now, in this 70-key process of clearing all the horizontal keys, it is evident that the orientation of a key is reversed once for each horizontal key in its row and column. Thus we have the following general procedure for opening the safe:

for each key, determine the total number H of horizontal keys in its row and column, and if H is odd, then turn the key, and if H is even, pass on to the next key without turning it.

This procedure of turning just the "odd" keys will certainly open the safe. But is there possibly a more efficient way of doing it? Thus let us consider a pretty argument which shows that our procedure takes us precisely to the minimal combination.

For each cell (i.e., key) let us call the parity of H the "parity of the cell." Now we need to make two observations, one obvious and the other surprising.

(a) in order to open the safe, the parity of every cell must be even (with only vertical keys at the end, H must be zero in every case),

(b) the only way to change the parity of a cell is to turn the key in that cell (surprisingly, turning other keys can't accomplish it).

Because of this, as you go from cell to cell in opening the safe, while turning an odd key switches the orientation of various keys, it changes the *parity* only of itself; the other odd keys are still in the same places awaiting their turns. We can establish (b) as follows.

Let k be a key on a safe of dimensions $2n \times 2n$. To reiterate, the parity of k is the number H of keys in k's row and column that have a horizontal orientation. Thus, the reversal of the orientation of a single key in k's row and column causes H to go up or down by one and therefore changes the parity of k. Similarly, each additional reversal toggles the parity and it follows that the reversal of an *odd* number of keys in k's row and column is required to achieve a change in the parity of k.

Consequently, turning a key t that is not in k's row or column fails to change the parity of k since it reverses a single key in k's row and a single key in its column, for an even total of two reversals (Figure 6a). Again, (Figure 6b), turning a key w that *is* in k's row or column, but isn't k itself, fails to change the parity of k since it reverses either just the $2n$ keys in its row or just the $2n$ keys in its column. However, turning k itself reverses all $4n - 1$ keys in its row and column, thus changing its parity.

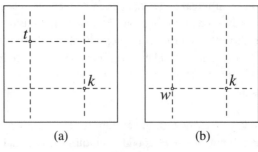

(a)   (b)

FIGURE 6

Recalling that the safe won't open unless all the cells have even parity, it follows that each key of odd parity *must* be turned in opening the safe, implying that every step of our procedure is necessary. Moreover, turning a key of even parity is a retrograde step, for that changes the parity of the key to odd, thus requiring a nullifying turn to bring it back to even parity. Hence our procedure calls for turning precisely those keys that must be turned, thus generating the combination exactly.

Since the combination can't contain a repeated key, one can never be required to turn more than all $4n^2$ keys on a $2n$ by $2n$ safe. Indeed, the question arises whether it is ever actually necessary to turn all the keys. However, if each key is horizontal at the start, then, for each key, H is the odd number $4n - 1$, implying that every key must be turned. Thus, if not more than 2002 turns are allowed, we can only guarantee to open a safe for which

$$4n^2 \leq 2002, \quad \text{that is, for } 2n \leq 44.$$

2. Out of the Mouths of Babes

Here is a letter Professor Hahn received from a school teacher friend. I have taken liberties with the text in order to tell the story.

> Dear Dr. Hahn,
>
> I suppose that you remember a puzzle I passed along a while back which was to take a 5×5 array of 25 lattice points and find a set of circles which passed through each point at least once (Figure 7).
>
> As is the privilege of teachers, I got to haul this problem out once again for this year's sixth graders. Several of them found

FIGURE 7

solutions using six circles (not hard), so I encouraged them to do better, and to find a five circle solution.

Figure 8 shows how five circles with center at the midpoint of the array pick up the points from the outside in. Unfortunately this doesn't get the center point itself.

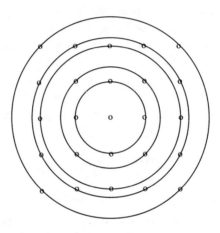

FIGURE 8

Eddie came to class the next day, face beaming, and announced that he had found a solution. He showed me his paper (Figure 9) which had three circles that hit every point but those on the diagonals, and two additional crossing lines along the diagonals.

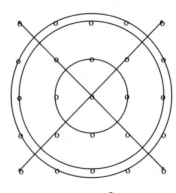

FIGURE 9

I said, "**Circles**, Eddie, **circles**!" He said, "Yes, the diagonals **are** circles, with infinite diameters!"

"Ok, ok," I said, "But there are solutions with five circles with *finite* diameters. See if you can find one of those."

Next day, Eddie is back, beaming again. He shows me his paper (Figure 10). This time it's five parallel lines running through the five rows of points. I say, "No, Eddie, circles with **finite** diameters!"

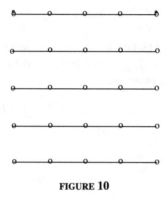

FIGURE 10

"They are!" he exclaims, "Look!" And he proceeds to roll the paper into a tube, turning the five lines into five circles wrapped around a cylinder (Figure 11).

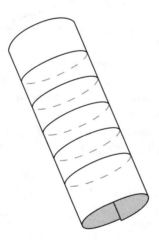

FIGURE 11

Eddie's poblem is posed in the exercises at the end of the essays and a solution is duly given in their solutions.

3. Pedal Triangles Revisited

Recall that the triangle DEF determined by the feet of the perpendiculars to the sides of $\triangle ABC$, extended if necessary, from a point P in the plane of ABC is called the pedal triangle of the point P with respect to $\triangle ABC$.

The following problem, due to Murray Klamkin (University of Alberta, retired), was posed in *From Erdős to Kiev* (page 235) with a nice solution by George Tsintsifas (Thessaloniki).

Let $\triangle DEF$ be the pedal triangle of an arbitrary point P inside $\triangle ABC$ (Figure 12). Let x, y, z, respectively, be the distances from P to the vertices A, B, C. Prove that

$$x^2 \sin 2A + y^2 \sin 2B + z^2 \sin 2C + 8\triangle DEF = 4\triangle ABC.$$

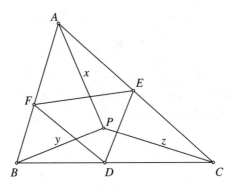

FIGURE 12

Now for the solution by Professor Liong-shin Hahn. As Erdős would say "it's one from the book!"

Dr. Hahn's brilliant move is to reflect P in each side of $\triangle ABC$ to get U, V, W, as shown in Figure 13.

Clearly, then,

$$AW = AV = x, \quad BW = BU = y, \quad \text{and} \quad CU = CV = z.$$

Also, it is clear that

$$\angle WAV = 2A, \quad \angle WBU = 2B, \quad \text{and} \quad \angle UCV = 2C.$$

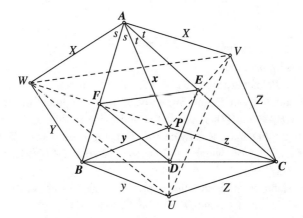

Consequently, the area of

$$\triangle AWV = \tfrac{1}{2}x^2 \sin 2A,$$

and similarly

$$\triangle BWU = \tfrac{1}{2}y^2 \sin 2B, \quad \triangle CUV = \tfrac{1}{2}z^2 \sin 2C.$$

That is to say,

$$x^2 \sin 2A = 2\triangle AWV, \quad y^2 \sin 2B = 2\triangle BWU, \quad z^2 \sin 2C = 2\triangle CUV.$$

Now, if we add these together and throw in $2\triangle UVW$ we get twice the entire figure $AWBUCV$:

$$x^2 \sin 2A + y^2 \sin 2B + z^2 \sin 2C + 2\triangle UVW = 2AWBUCV.$$

But, clearly, the reflections unfold $\triangle ABC$ and result in a figure of twice the area, and so we have

$$AWBUCV = 2\triangle ABC,$$

and therefore

$$x^2 \sin 2A + y^2 \sin 2B + z^2 \sin 2C + 2\triangle UVW = 4\triangle ABC.$$

Thus it remains only to show that

$$2\triangle UVW = 8\triangle DEF,$$

that is,

$$\triangle UVW = 4\triangle DEF.$$

But this is immediate.

Since *PF* is perpendicular to *AB*, *PF* reflects into *WF*, making *PFW* straight and *F* its midpoint. Similarly, *E* is the midpoint of *PV*, and we have in $\triangle PWV$ that *WV* is parallel to and twice as long as *FE*. Similarly, *UW* is twice *FD* and *UV* is twice *DE*. Thus the sides of $\triangle UVW$ are twice those of $\triangle DEF$, from which it follows that the area of $\triangle UVW$ is four times that of $\triangle DEF$.

Dr. Hahn observes that this approach is successful even with obtuse triangles *ABC* and for points *P* outside the triangle.

4. Finally let us look again at Problem 4, page 93, of *In Pólya's Footsteps*. I have always felt a little uneasy with the published solution, which is long and involved and not all that exciting. The problem originally appeared on the 1989 AIME examination.

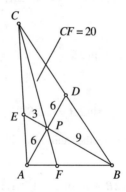

FIGURE 14

Referring to Figure 14, in $\triangle ABC$, *AD*, *BE*, and *CF* intersect at *P* such that

$$AP = PD = 6, EP = 3, PB = 9, \quad \text{and} \quad CF = 20.$$

What is the area of $\triangle ABC$?

A cevian is a segment that goes from a vertex of a triangle to an interior or exterior point in the opposite side. The following two complementary theorems are often useful when dealing with three concurrent cevians like *AD*, *BE*, and *CF*.

Theorem 1.

$$\frac{PD}{AD} + \frac{PE}{BE} + \frac{PF}{CF} = 1$$

Theorem 2.

$$\frac{PA}{AD} + \frac{PB}{BE} + \frac{PC}{CF} = 2.$$

The simple proofs of these theorems are given at the end of this part so that we might get on with the solution without delay. We note that Theorem 2 is not used in the solution.

By Theorem 1, then, we have

$$\frac{6}{12} + \frac{3}{12} + \frac{PF}{20} = 1, \quad \text{giving } PF = 5,$$

and consequently, $CP = 15$.

Let *PA*, *PB*, and *PC* partition $\triangle ABC$ into triangles of areas u, v, and w, as in Figure 15.

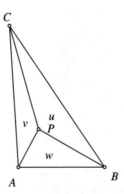

FIGURE 15

Now, the areas of triangles having equal altitudes are proportional to the lengths of the corresponding bases. Since $PF = 5$ and $CP = 15$, then

$$\triangle PAF = \tfrac{1}{3}v \quad \text{and} \quad \triangle PBF = \tfrac{1}{3}u.$$

Adding we get $w = \frac{1}{3}(v + u)$. Similarly, since

$$EP = \tfrac{1}{3}PB, \quad \triangle PCE = \tfrac{1}{3}u, \quad \text{and} \quad \triangle PEA = \tfrac{1}{3}w,$$

giving $v = \frac{1}{3}(u + w)$. Solving, we have

$$3w = v + u \quad \text{and} \quad 3v = u + w.$$

Subtracting we get

$$3(w - v) = v - w,$$
$$4(w - v) = 0,$$

and

$$w = v.$$

Then

$$3w = v + u \quad \text{gives } u = 2w = 2v.$$

Therefore

$$\frac{AF}{FB} = \frac{\triangle PAF}{\triangle PFB} = \frac{\frac{1}{3}v}{\frac{1}{3}u} = \frac{v}{u} = \frac{1}{2}.$$

That is to say, BF is twice FA.

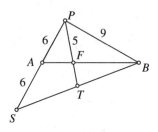

FIGURE 16

Now Professor Hahn makes the brilliant move of extending PA its own length to S (Figure 16), thus making BA a median of $\triangle PSB$, and since $BF = 2FA$, then F is in fact the centroid of the triangle. Hence PFT is also a median and therefore F also divides PT in the ratio of 2 to 1, and we have

$$PT = \tfrac{3}{2} \cdot PF = \tfrac{3}{2} \cdot 5 = \tfrac{15}{2}.$$

Now, there is a theorem that says the sum of the squares of two sides of a triangle is equal to twice the square of the median to the third side plus one-half the square of the third side: in Figure 17,

$$a^2 + b^2 = 2m^2 + \tfrac{1}{2}c^2.$$

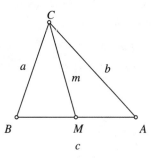

FIGURE 17

(This is immediately established by applying the law of cosines to $\triangle MBC$ to get a^2 and to $\triangle MAC$ to get b^2 and adding the results.) Thus it follows in Figure 16 that

$$SP^2 + PB^2 = 2PT^2 + \tfrac{1}{2}SB^2,$$
$$12^2 + 9^2 = 2\left(\tfrac{15}{2}\right)^2 + \tfrac{1}{2}SB^2,$$
$$225 = \tfrac{1}{2} \cdot 225 + \tfrac{1}{2}SB^2,$$

giving

$$SB^2 = 225.$$

In triangle SPB, then, we have

$$SP^2 + PB^2 = 12^2 + 9^2 = 225 = SB^2,$$

implying that angle SPB is a right angle by the converse of the theorem of Pythagoras. The area w of $\triangle PAB$ is therefore

$$w = \tfrac{1}{2} \cdot \text{ base } PA \cdot \text{ height } PB = \tfrac{1}{2} \cdot 6 \cdot 9 = 27.$$

Thus

$$v = w = 27 \quad \text{and} \quad u = 2w = 54,$$

making

$$\triangle ABC = u + v + w$$
$$= 54 + 27 + 27 = 108.$$

Proof of Theorem 1:

$$\frac{PD}{AD} + \frac{PE}{BE} + \frac{PF}{CF} = 1.$$

Let PL and CN be perpendicular to AB (Figure 18). Then right triangles PLF and CNF are similar and

$$\frac{PF}{CF} = \frac{PL}{CN} = \frac{\left(\frac{1}{2} \cdot AB\right) \cdot PL}{\left(\frac{1}{2} \cdot AB\right) \cdot CN} = \frac{\triangle PAB}{\triangle ABC}.$$

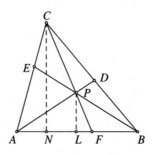

FIGURE 18

Similarly

$$\frac{PD}{AD} = \frac{\triangle PBC}{\triangle ABC} \quad \text{and} \quad \frac{PE}{BE} = \frac{\triangle PCA}{\triangle ABC}.$$

Adding we get

$$\frac{PD}{AD} + \frac{PE}{BE} + \frac{PF}{CF} = \frac{\triangle PBC + \triangle PCA + \triangle PAB}{\triangle ABC}$$

$$= \frac{\triangle ABC}{\triangle ABC} = 1.$$

Theorem 2.

$$\frac{PA}{AD} + \frac{PB}{BE} + \frac{PC}{CF} = 2.$$

It is readily seen that Theorem 2 is just a corollary of Theorem 1: since

$$\frac{PA}{AD} + \frac{PD}{AD} = \frac{AD}{AD} = 1,$$

then

$$\frac{PA}{AD} = 1 - \frac{PD}{AD}, \text{ etc.,}$$

giving

$$\frac{PA}{AD} + \frac{PB}{BE} + \frac{PC}{CF} = \left(1 - \frac{PD}{AD}\right) + \left(1 - \frac{PE}{BE}\right) + \left(1 - \frac{PF}{CF}\right)$$

$$= 3 - \left(\frac{PD}{AD} + \frac{PE}{BE} + \frac{PF}{CF}\right) = 3 - 1 = 2.$$

From the 2002 New Mexico Mathematics Contest

A Problem from the First Round of the New Mexico Mathematics Contest (November, 2002) (Slightly reworded)

(a) What is the radius of the smallest circle that will cover any triangle of unit perimeter?

(b) What is the radius of the smallest circle that will cover any planar 2002-gon P of unit perimeter?

Solution 1

Since the sides of a *degenerate* planar polygon of unit perimeter "double cover" a segment of length $\frac{1}{2}$, a circle K of diameter $\frac{1}{2}$ suffices for the degenerate cases of both parts (a) and (b). Let us see whether K is also sufficient for the nondegenerate cases.

(a) Let ABC be a nondegenerate triangle of unit perimeter. If BC is a longest side, then BC must be less than $\frac{1}{2}$ in order to prevent the entire perimeter from exceeding unity. Let K be placed on the triangle so that a longest side BC is a chord (BC is less than the diameter) and A is on the same side of BC as the center O of K (Figure 1). We wish to show that A lies in K.

Let BQ and CS be perpendicular to BC. Then vertex A must lie strictly between BQ and CS in order to avoid the contradiction of having either AB or AC greater than the longest side BC. Furthermore, A must occur on one side or the other of the line of the central diameter NOM which is perpendicular to BC and which bisects BC at M. For definiteness, suppose A is in the strip between the lines NM and CS, possibly on the line NM. Also, let CS cross K at T and let P be a point on the minor arc NT.

Let us take this opportunity to apply Archimedes' Theorem of the Broken Chord (Figure 1a):

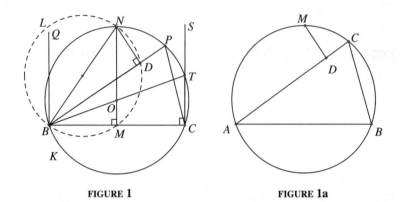

FIGURE 1   FIGURE 1a

if M is the midpoint of circular arc ACB, then the foot D of the perpendicular from M to the longer of the chords AC and BC bisects the polygonal path $AC + CB$ (a proof is given in the Appendix to this section).

Accordingly, the foot D of the perpendicular to BP from N bisects the polygonal path $BP + PC$. Thus the perimeter of $\triangle BPC$ is $BC + 2BD$. Now, D lies on the circle L on diameter BN, and as P moves along its arc toward T, D slides around L toward M (which also lies on L), shortening the chord BD in L as it does so. It follows that the perimeter of $\triangle BPC$ *decreases* as P approaches T along the arc and that the perimeter of $\triangle BTC$ is strictly less than the perimeter of any of the triangles BPC.

Now, since $\angle BCT$ is a right angle, BT is a diameter of K and therefore of length $\frac{1}{2}$. Since $BC + CT > BT$ by the triangle inequality, the perimeter of $\triangle BTC$ must exceed 1. And since $\triangle BPC$ has even greater perimeter, the perimeter of $\triangle BPC$ also exceeds 1, and it follows that A cannot lie on the arc NT.

Similarly, A cannot lie on the arc on the other side of N and we conclude that A cannot lie anywhere on the circumference of K.

Now, if A were to lie outside K (Figure 2), then, letting AB cross K at P,

$$\text{the perimeter of } \triangle ABC = AB + AC + BC$$
$$= BP + (PA + AC) + BC$$
$$> BP + PC + BC$$
$$= \text{the perimeter of } \triangle BPC$$
$$> 1,$$

a contradiction. Hence A must lie strictly inside K.

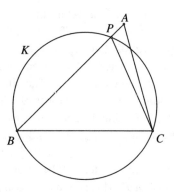

FIGURE 2

It is interesting that K always provides a "*loose fit*" in the nondegener-ate case, even though no smaller circle can provide coverage in all cases (no smaller circle can cover all triangles with sides $\frac{1}{2} - x$, $\frac{1}{2} - x$, and $2x$, as x approaches zero).

Comment   Another neat way of seeing that the perimeter of $\triangle BPC$ increases as P moves along arc TN from T is provided by ellipses having foci at B and C. Let such an ellipse E be constructed to pass through T by keeping taut a loop of string around pins at B and C (Figure 3).

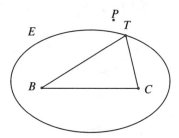

FIGURE 3

Then the perimeter of triangle BCT is just the length of the loop used in drawing E. Now, moving from T to a point P along the arc TN results in a point *outside* E. The perimeter of $\triangle BPC$ is the length of the loop needed to consruct the ellipse with foci at B and C that passes through P. Clearly a longer loop is needed to make a larger confocal ellipse, and it follows that the perimeter of $\triangle BCP$ increases as P moves along the arc away from T.

You might find this alternative approach quicker and easier than using the Theorem of the Broken Chord, but it is such a red-letter day when one happens upon an application of Archimedes' marvelous theorem that there was never a shadow of a doubt which approach I would feature in the solution.

Now let us consider the 2002-gon P.

(b) The section of the perimeter of P between vertices P_i and P_j cannot have length less than the segment $P_i P_j$ itself. Thus the perimeter of any triangle $P_i P_j P_k$ that is determined by three vertices of P cannot have perimeter exceeding unity without also forcing the perimeter of P beyond unity. For the same reason, no side or diagonal $P_i P_j$ of our nondegenerate polygon P could be as long as $\frac{1}{2}$.

Now let a large circle C that encloses P be shrunk until a tight contact is made with outlying vertices of P. There are only two kinds of contact that can prevent further shrinking:

(i) C makes contact with two vertices P_i and P_j at the ends of a diameter (Figure 4(i)),

(ii) C contacts the three vertices P_i, P_j, P_k of an *acute-angled* triangle (Figure 4(ii)).

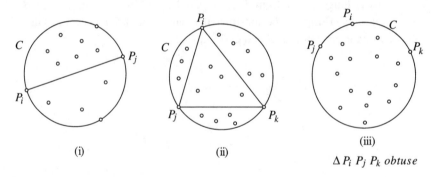

(i)   (ii)

(iii)

$\Delta P_i P_j P_k$ *obtuse*

FIGURE 4

(Observe that if C makes contact only at the vertices of an obtuse-angled triangle, it is not sufficient to stop the shrinking: since the vertices must lie in the interior of a semicircle of C, C can be pushed away from them (perhaps only infinitesimally) to allow further shrinking (Figure 4(iii)).)

In case (i), since the side or diagonal $P_i P_j$ is less than $\frac{1}{2}$, C is actually smaller than our circle K, implying K is more than sufficient to cover P.

In case (ii), the perimeter of $\Delta P_i P_j P_k$ cannot exceed unity, and hence it follows by part (a) that $\Delta P_i P_j P_k$, and simultaneously the circle C and the polygon P, can be covered by the circle K.

Since the number of sides of P has not been a consideration, we may conclude that K is a universal cover of all planar n-gons of unit perimeter and that, except in the degenerate cases, it is always a loose fit.

Solution 2

Now let us return to part (a) for the wonderful solution by Professor V. Koltchinskii, University of New Mexico, who proposed the problem for the contest.

(a) Let ABC be a nondegenerate triangle of unit perimeter. Thus the length of each side is less than $\frac{1}{2}$, in particular AB and AC. Therefore, going halfway around the perimeter from A would carry one beyond vertex B in the one direction, and beyond C in the other, to a point X in the interior of side BC (Figure 5). Thus

$$AB + BX = AC + CX = \tfrac{1}{2}.$$

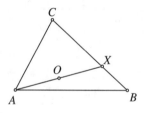

FIGURE 5

It follows from the triangle inequality that

$$AX < AB + BX = \tfrac{1}{2},$$

implying that the distance from A to the midpoint O of AX is less than $\frac{1}{4}$. Thus the circle K with center O and radius $\frac{1}{4}$ certainly covers the point A.

But K also covers vertex B! Professor Koltchinskii proves this brilliantly by completing parallelogram $ABXD$ (Figure 6).

Then opposite sides DX and AB are equal, and since O is the midpoint of diagonal AX, it is also the midpoint of diagonal DB. Thus

$$OB = \tfrac{1}{2}DB < \tfrac{1}{2}(DX + XB) \quad \text{(by the triangle inequality)},$$
$$= \tfrac{1}{2}(AB + BX) = \tfrac{1}{2}\left(\tfrac{1}{2}\right) = \tfrac{1}{4}.$$

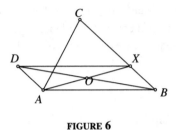

FIGURE 6

Similarly, by completing parallelogram $ACXE$ (Figure 7), it follows that $OC < \frac{1}{4}$ and therefore the circle K covers the entire triangle ABC.

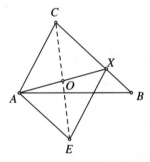

FIGURE 7

(b) Although Professor Koltchinskii's approach also suffices for part (b), let us close with a beautiful solution by Professor Liong-shin Hahn, University of New Mexico, retired.

We have seen that a circle of radius $\frac{1}{4}$ suffices for the degenerate case. Therefore let P be a nondegenerate planar 2002-gon $P_1 P_2 \ldots P_{2002}$ of unit perimeter. Now Professor Hahn draws 2002 circles of radius $\frac{1}{4}$, one around each vertex P_i as center, and then proceeds to a quick resolution of the problem with an application of Helly's theorem.

As we have seen earlier, the triangle $P_i P_j P_k$ formed by three vertices of P must have perimeter not exceeding unity and therefore, by part (a), the triangle can be covered by a circle K of radius $\frac{1}{4}$. Hence the distance from the center O of K to each vertex cannot exceed the radius $\frac{1}{4}$. Thus our circle of radius $\frac{1}{4}$ which has center at P_i covers the point O. Similarly, the circles at P_j and P_k also cover O, and it follows that each subset of three of our 2002 circles have a point in common (Figure 8).

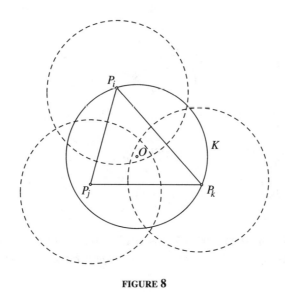

FIGURE 8

Now, Helly's theorem (which we will prove shortly) says that if each three of a collection of n convex sets have a point in common, then there is a point that is common to all n of the sets (a convex set is one with the property that, for any two points A and B of the set, the entire segment AB belongs to the set). Since circles are convex sets, it follows that there is a point X that lies in all 2002 of our circles. Hence for all vertices P_i, the distance XP_i must not exceed the radius $\frac{1}{4}$ of the circles, from which it follows that a circle of radius $\frac{1}{4}$ with center X covers all 2002 vertices of P.

Helly's Theorem

Before going on with the proof of Helly's remarkable theorem, let us note two basic properties of convex sets.

(i) If a convex set S contains three points A, B, C, then it contains the entire triangle ABC. (Clearly the sides are in S, and so is any interior point P since it lies on a segment AD which joins two points of S (Figure 9).)

(ii) The intersection of a collection of convex sets is a convex set. (If A and B are two points of the intersection, then A and B belong to each of the sets, and since they are convex, each set also contains the entire segment AB. Hence AB is in the intersection, implying the intersection is also a convex set.)

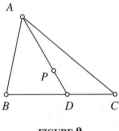

FIGURE 9

Now to the proof of Helly's theorem.

Helly's Theorem (1923). *If each three of a collection of n convex sets have a point in common, there is a point that is common to all n of the sets.*

1. We begin by proving the result for a collection of four convex sets, φ_1, φ_2, φ_3, φ_4, each three of which have a point in common. Let A_i be a point that is common to the trio of sets which omits φ_i. Thus A_1 lies in φ_2, φ_3, φ_4, A_2 in φ_1, φ_3, φ_4, and so on. Now, four points can give rise to only two configurations in the plane—either one is contained in the triangle formed by the other three, or they determine a convex quadrilateral. Suppose, for definiteness, that, in each case, A_1, A_2, A_3, A_4 occur as labelled in Figure 10.

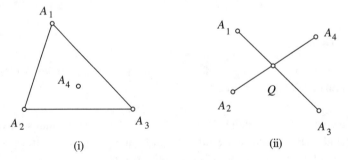

FIGURE 10

In case (i), each vertex of the triangle lies in φ_4, and since φ_4 is convex, it contains the entire triangle. Hence A_4 also lies in φ_4, and it follows that A_4 lies in all four of the sets.

In case (ii), let $A_1 A_3$ intersect $A_2 A_4$ at Q. Now, φ_4 contains the entire triangle $A_1 A_2 A_3$, and therefore also the point Q on side $A_1 A_3$. Similarly, Q lies in each of the four sets, and the conclusion follows.

2. Proceeding by induction, suppose the result holds for a collection of n sets, $n \geq 4$, each three of which have a point in common, and that S is a collection of $n + 1$ such sets $\varphi_1, \varphi_2, \ldots, \varphi_n, \varphi_{n+1}$. Since each three of these $n + 1$ sets have a point in common, then, for distinct i and j less than n, each three of the four sets $\varphi_i, \varphi_j, \varphi_n, \varphi_{n+1}$ have a point in common. By step 1, these four sets have a common point X, which, belonging to both φ_n and φ_{n+1}, also belongs to their intersection φ. Thus, for each choice of i and j, there is a point X that is common to the three convex sets $\varphi_i, \varphi_j, \varphi$, and it follows that each three of the n convex sets $\varphi_1, \varphi_2, \ldots, \varphi_{n-1}, \varphi$ have a point in common. By the induction hypothesis, then, there is a point Y that is common to all n of these sets, and, belonging to φ, Y belongs to each of φ_n and φ_{n+1}, implying that it lies in all $n + 1$ of the sets of S.

Thus ends our story of Professor Koltchinskii's wonderful problem.

Appendix

Archimedes' Theorem of the Broken Chord. *If M is the midpoint of circular arc ACB, then the foot D of the perpendicular from M to the longer of the chords AC and BC bisects the polygonal path $AC + CB$ (Figure 11).*

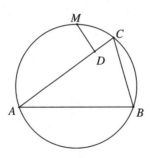

FIGURE 11

Proof: Let AC be extended to F so that $CF = CB$ (Figure 12). We would like to show that the foot D is the midpoint of AF.

Clearly, $\triangle BCF$ is isosceles with equal angles x at B and F. Thus the exterior $\angle ACB = 2x$. Since $\angle ACB = \angle AMB$ in the same segment of the circle, we have $\angle AMB = 2x$.

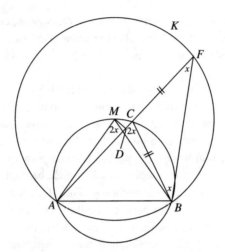

FIGURE 12

Now, let K be the circle through A, B, and F. Then the center of K is the point on the perpendicular bisector of AB at which AB subtends an angle that is twice the angle x which AB subtends at F on the circumference. That is to say, the center must be M, and since the perpendicular from the center of a circle to a chord bisects the chord, it follows that D is indeed the midpoint of AF.

Two Solutions by Achilleas Sinefakopoulos

Achilleas Sinefakopoulos is from Larissa, Greece, and is currently (2003) a Ph.D. candidate at Cornell University. He very kindly sent me this sample of his work which he has allowed me to present in my own words.

1. First we have Achilleas' fine solution of Problem 1465 from *Mathematics Magazine* (February, 1996, page 69), which was proposed by Stephen W. Knox, University of Illinois, Urbana, Illinois.

> In every 2-coloring of the plane, prove that there are monochromatic triangles of every conceivable shape. That is to say, in every 2-coloring of the plane, prove there is a monochromatic triangle that is similar to any given triangle *PQR*.

Let each point of the plane be colored either red or blue. We wish to show that no coloring can fail to create a monochromatic triangle that is similar to $\triangle PQR$.

Now, every triangle has at least two acute angles. Suppose $\angle P$ is one of the acute angles in $\triangle PQR$. We begin by constructing a triangle $\triangle ABC$ as follows. Vertices B and C are taken at any two points which have the same color. At B and C, angles Q and R are drawn to complete the triangle (Figure 1). Thus $\triangle ABC$ is

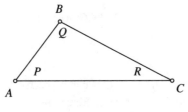

FIGURE 1

(i) similar to $\triangle PQR$,

(ii) has an acute angle at A, namely $\angle P$,

(iii) has vertices B and C the same color.

Clearly, a triangle that is similar to $\triangle ABC$ is also similar to $\triangle PQR$.

Since angle A is acute, we can proceed to construct a network of triangles that are similar to $\triangle ABC$ of the kind that is shown in Figure 2:

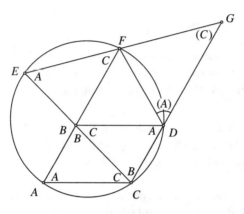

FIGURE 2

We begin by completing parallelogram $ABDC$ and drawing the circumcircle of $\triangle ACD$. Let AB and CB meet the circle at F and E, respectively. Then, in the parallelogram, $\angle BCD = B$, and in the same segment of the circle, angle $E = A$. Since the sum of angles A and B at E and C is less than a straight angle, Euclid's parallel postulate implies that EF and CD meet at some point G on the same side of EC as F and D.

Now, the angles of $\triangle ABC$ occur in several places in the figure and it is not difficult to show that each of the five triangles

$$BCD, EFB, ECG, AED, FDG,$$

is similar to $\triangle ABC$:

In the parallelogram, A, B, C occur, respectively, at D, C, and B, and so $\triangle BCD$ is similar to $\triangle ABC$. Since angles in the same segment of a circle are equal, angles A and C occur again at E and F, making

$\triangle EFB$ similar to $\triangle ABC$. Then, too, $\triangle ECG$, with angles A and B at E and C, is similar to $\triangle ABC$, and hence has angle C at G.

Furthermore, as shown in Figure 3, at C on the circumference, ED subtends angle B and EA subtends angle C. Thus angles B and C occur again at A and D, making $\triangle AED$ similar $\triangle ABC$.

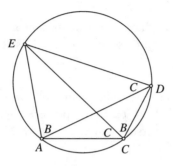

FIGURE 3

Going back to Figure 2, since $FACD$ is cyclic, the exterior angle at D is equal to the interior angle at the opposite vertex A, that is, $\angle FDG = A$, which, with angle C at G, makes $\triangle FDG$ similar to $\triangle ABC$.

Finally, let us add color to this network by labeling the vertices r for red and b for blue (Figure 4). Recall that B and C were chosen to have the same color, say blue. Now, since at least two vertices of a triangle must be

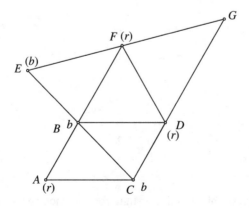

FIGURE 4

the same color, a monochromatic triangle can only be avoided by using a different color on the third vertex. Therefore a monochromatic triangle that is similar to $\triangle ABC$ will have to be formed unless

1. A and D are red (in triangles $\triangle ABC$ and BCD),
2. E is blue (in $\triangle AED$),
3. F is red (in $\triangle EFB$).

At G, then, we see that there is no avoiding a monochromatic triangle that is similar to $\triangle ABC$, for if G is red, then $\triangle FDG$ is all red, and if G is blue, $\triangle ECG$ is all blue.

2. Now we turn to Achilleas' splendid solution of Problem 1909 in *Crux Mathematicorum* (Vol. 20, 1994, page 295) proposed by Charles R. Diminnie, Saint Bonaventure University, Saint Bonaventure, New York:

Solve the recurrence

$$p_{n+1} = 5p_n(5p_n^4 - 5p_n^2 + 1), \quad p_0 = 1.$$

It is easy to calculate that

$$p_1 = 5(5 - 5 + 1) = 5,$$
$$p_2 = 25(3125 - 125 + 1) = 25(3001) = 75025,$$
$$p_3 = 5 \cdot 75025(5 \cdot 75025^4 - 5 \cdot 75025^2 + 1),$$

which is enormous. This is getting us nowhere fast!

Achilleas takes us by surprise at the very beginning by introducing the Chebyshev polynomials, which are defined recursively by

$$f_n(x) = x \cdot f_{n-1}(x) - f_{n-2}(x).$$

With initial cases

$$f_0(x) = 2 \quad \text{and} \quad f_1(x) = x,$$

we have

$$f_2(x) = x^2 - 2,$$
$$f_3(x) = x^3 - 3x,$$
$$f_4(x) = x^4 - 3x^2 - (x^2 - 2) = x^4 - 4x^2 + 2,$$
$$f_5(x) = x^5 - 4x^3 + 2x - (x^3 - 3x) = x^5 - 5x^3 + 5x,$$
$$- - - - - - - - - - - - - - - -.$$

Now, a sharp eye might notice a slight resemblance between the given recursion,

$$p_{n+1} = 5p_n(5p_n^4 - 5p_n^2 + 1) = 5(5p_n^5 - 5p_n^3 + p_n)$$

and

$$f_5(x) = x^5 - 5x^3 + 5x.$$

In fact,

$$\begin{aligned} f_5(\sqrt{5}p_n) &= (\sqrt{5}p_n)^5 - 5(\sqrt{5}p_n)^3 + 5(\sqrt{5}p_n) \\ &= 5\sqrt{5}(5p_n^5 - 5p_n^3 + p_n) \\ &= \sqrt{5}p_{n+1}. \end{aligned}$$

Hence the given recurrence is

$$p_{n+1} = \tfrac{1}{\sqrt{5}} f_5(\sqrt{5}p_n), \quad p_0 = 1.$$

Now, it is just as easy to solve the entire family of recurrences

$$p_{n+1} = \tfrac{1}{\sqrt{5}} f_m(\sqrt{5}p_n), \quad \text{where } m \text{ is any odd number} > 1,$$

as it is to solve the individual case of $m = 5$. We can only wonder what possessed Achilleas to connect p_n with the Fibonacci sequence $\{F_n\}$ and the Chebychev polynomials. Granted there is the resemblance that we already mentioned, but it is still very impressive to come up with the conjecture that

$$p_n = F_{m^n}.$$

Once proposed, however, this is not that difficult to confirm, thanks to a couple of nice properties of the Chebyshev polynomials and the conjugate irrationals in Binet's formula for the Fibonacci numbers.

Recall that Binet's formula is

$$F_n = \frac{\alpha^n - \beta^n}{\alpha - \beta},$$

where α and β are the roots of $x^2 - x - 1 = 0$:

$$\alpha = \frac{1 + \sqrt{5}}{2} \quad \text{and} \quad \beta = \frac{1 - \sqrt{5}}{2}.$$

Clearly $\alpha - \beta = \sqrt{5}$, and since the product of the roots $\alpha\beta = -1$, then $\beta = \frac{-1}{\alpha}$ and

$$F_n = \frac{1}{\sqrt{5}} \left(\alpha^n - \frac{(-1)^n}{\alpha^n} \right).$$

For n odd, then,

$$F_n = \frac{1}{\sqrt{5}} \left(\alpha^n + \frac{1}{\alpha^n} \right).$$

Since m is odd, so is m^n, and our conjecture is that

$$p_n = F_{m^n} = \frac{1}{\sqrt{5}} \left(\alpha^{m^n} + \frac{1}{\alpha^{m^n}} \right).$$

This just suits the Chebyshev polynomials beautifully, for one of their important properties is that

$$f_n \left(y + \frac{1}{y} \right) = y^n + \frac{1}{y^n},$$

the proof of which is left as an easy exercise in induction. Recall that we are trying to solve the recursion

$$p_{n+1} = \frac{1}{\sqrt{5}} f_m(\sqrt{5} p_n), \quad p_0 = 1.$$

Proceeding by induction, suppose, for some value of n, that

$$p_n = F_{m^n} = \frac{1}{\sqrt{5}} \left(\alpha^{m^n} + \frac{1}{\alpha^{m^n}} \right).$$

Then

$$\sqrt{5} p_n = \alpha^{m^n} + \frac{1}{\alpha^{m^n}},$$

and from the recursion, we have

$$p_{n+1} = \frac{1}{\sqrt{5}} f_m \left(\alpha^{m^n} + \frac{1}{\alpha^{m^n}} \right).$$

Using $y = \alpha^{m^n}$ in the Chebyshev property, we get

$$p_{n+1} = \frac{1}{\sqrt{5}} \left[(\alpha^{m^n})^m + \frac{1}{(\alpha^{m^n})^m} \right],$$

$$= \frac{1}{\sqrt{5}} \left(\alpha^{m^{n+1}} + \frac{1}{\alpha^{m^{n+1}}} \right)$$

$$= F_{m^{n+1}} \quad \text{since } m^{n+1} \text{ is odd.}$$

Since $p_0 = 1 = F_1 = F_{m^0}$, it follows that $p_n = F_{m^n}$ for all $n \geq 0$.

Recalling that $m = 5$ in the given recurrence, its solution is therefore $p_n = F_{5^n}$.

Alternative Solutions by George Evagelopoulos to Three Problems from the 1982 West German Olympiad

1. A set of numbers is called "sum-free" if no two of them, the same or different, add up to a member of the set.

> What is the maximum size of a sum-free subset A of
> $\{1, 2, 3, \ldots, 2n + 1\}$?

If the greatest element in A is an odd number $2k + 1$, A would be a subset of $\{1, 2, 3, \ldots, 2k + 1\}$. The first $2k$ of these integers go together into k pairs with sum $2k + 1$:

$$(1, 2k), (2, 2k - 1), (3, 2k - 2), \ldots, (k, k + 1).$$

In order to be sum-free, A could not have more than one integer from each pair, and therefore, not forgetting to count $2k + 1$, its maximum size could not exceed $k + 1$. Since k is not greater than n, A could not have more than $n + 1$ elements.

On the other hand, suppose the greatest integer in A is an even number $2k$, making A a subset of $\{1, 2, 3, \ldots, 2k\}$. Then, since $k + k = 2k$, k could not belong to A, and the $2k - 2$ integers $\{1, 2, \ldots, k - 1, k + 1, \ldots, 2k - 1\}$ go into $k - 1$ pairs with sum $2k$:

$$(1, 2k - 1), (2, 2k - 2), \ldots, (k - 1, k + 1).$$

Thus the maximum size of A could not exceed $(k - 1) + 1 = k$, which is not greater than n.

In any case, then, the cardinality of A cannot exceed $n + 1$. But it is easy to see that there are sum-free subsets of size $n + 1$:

(a) the odd integers $\{1, 3, 5, \ldots, 2n+1\}$, since the sum of two odd numbers is even,

(b) the $n+1$ consecutive integers $\{n+1, n+2, \ldots, 2n+1\}$, since the sum of any two, the same or different, exceeds $2n + 1$.

Thus we conclude that the mazimum size of A is indeed $n + 1$.

2. What is the sum S of the greatest odd divisors of the numbers

$$1, 2, 3, \ldots, 2^n?$$

We observe that, since 2^n is even, its greatest odd divisor is 1, and therefore

$S = 1 + $ the sum of the greatest odd divisors of $1, 2, 3, \ldots, 2^n - 1$.

George approaches the problem directly by noting that the greatest odd divisor of a positive integer k is the odd number $2i - 1$ that results when k is divided by 2 as often as possible, that is, the number $2i - 1$, where $k = (2i - 1)2^t$.

Thus we seek the sum

$$S = 1 + \sum_{k=1}^{2^n-1} (2i - 1).$$

George's clever idea is to consider which *odd* numbers are associated with a specified power 2^t; that is,

"How far can the sequence $1 \cdot 2^t, 3 \cdot 2^t, 5 \cdot 2^t, \ldots, (2i-1)2^t$ be extended without going past $2^n - 1$?"

Clearly t can't be as big as n, and for all $t = 0, 1, 2, \ldots, n-1$, the condition

$$(2i - 1)2^t \leq 2^n - 1$$

is equivalent to

$$2i - 1 \leq 2^{n-t} - \frac{1}{2^t},$$

and, since $2i - 1$ is an integer, to

$$2i - 1 \leq 2^{n-t} - 1.$$

Therefore the sum of the greatest odd divisors of the *odd* multiples of 2^t is

$$S_t = 1 + 3 + 5 + \cdots + (2^{n-t} - 1)$$
$$= 1 + 3 + 5 + \cdots + [2(2^{n-t-1}) - 1],$$

an arithmetic series of 2^{n-t-1} terms. Hence

$$S_t = \frac{2^{n-t-1}}{2}[1 + (2^{n-t} - 1)] = (2^{n-t-1})^2 = 4^{n-t-1}.$$

We observe that the even multiples of 2^t are odd multiples of larger powers of 2. For example $6 \cdot 2^t = 3 \cdot 2^{t+1}$ is an odd multiple of 2^{t+1} and its greatest odd divisor 3 is contained in the sum S_{t+1}; similarly, $8 \cdot 2^t = 1 \cdot 2^{t+3}$ contributes its 1 to the sum S_{t+3}.

Thus the required sum is

$$S = 1 + \sum_{t=0}^{n-1} S_t$$

$$= 1 + \sum_{t=0}^{n-1} 4^{n-t-1}$$

$$= 1 + (4^{n-1} + 4^{n-2} + \cdots + 4 + 1)$$

$$= 1 + \frac{4^n - 1}{4 - 1}$$

$$= \frac{4^n + 2}{3}.$$

3. If a positive integer n makes $4^n + 2^n + 1 = p$, a prime number, prove that n must be a power of 3.

Let all the 3's be factored out of n to give $n = 3^k t$. Thus the resulting factor t is an integer which is at least 1 and which is not further divisible by 3. Hence, for some nonnegative integer s,

$$t = 3s + r, \quad \text{where } r = 1 \text{ or } 2.$$

Then

$$p = 4^{3^k t} + 2^{3^k t} + 1$$

$$= (2^{3^k})^{2t} + (2^{3^k})^t + 1.$$

We shall prove shortly that, because $t = 3s + r$, where $r = 1$ or 2, this expression contains the factor

$$F = (2^{3^k})^2 + 2^{3^k} + 1.$$

Under this assumption, because p is a prime number and F is greater than 1, F could only be p itself, and we have

$$p = (2^{3^k})^{2t} + (2^{3^k})^t + 1 = F = (2^{3^k})^2 + 2^{3^k} + 1.$$

Now, we know that $t \geq 1$, and if t were to exceed 1, each of the first two terms on the left side would be greater than the corresponding term on the right side, making equality impossible. Thus t must be 1, yielding the desired conclusion

$$n = 3^k t = 3^k, \quad \text{a power of 3.}$$

We conclude by showing that, for $t = 3s + r$ and $r = 1$ or 2,

$$Z = (2^{3^k})^{2t} + (2^{3^k})^t + 1 \quad \text{contains the factor} \quad F = (2^{3^k})^2 + 2^{3^k} + 1.$$

We shall consider the case $r = 1$ and leave the similar case $r = 2$ for the interested reader.

Letting $x = 2^{3^k}$, and noting that $t = 3s + 1$, we have

$$Z = (2^{3^k})^{2t} + (2^{3^k})^t + 1 = x^{6s+2} + x^{3s+1} + 1.$$

We want to show that $F = x^2 + x + 1$ is a factor of Z.

How clever of George to multiply by $x - 1$ to give

$$
\begin{aligned}
(x - 1)Z &= x^{6s+3} + x^{3s+2} + x - x^{6s+2} - x^{3s+1} - 1 \\
&= (x^{6s+3} - 1) - (x^{6s+2} - x^{3s+2}) - (x^{3s+1} - x) \\
&= [(x^3)^{2s+1} - 1] - x^{3s+2}[(x^3)^s - 1] - x[(x^3)^s - 1],
\end{aligned}
$$

where each term contains a factor of the form $(x^3)^m - 1$. Now, for all positive integers m,

$$(x^3)^m - 1 = (x^3 - 1)[(x^3)^{m-1} + (x^3)^{m-2} + \cdots + x^3 + 1]$$

and so $(x^3 - 1)$ a factor of $(x - 1)Z$. That is to say,

$$x^3 - 1 = (x - 1)(x^2 + x + 1)$$

is a factor of $(x - 1)Z$, and we conclude that $(x^2 + x + 1)$ is a factor of Z, completing the proof.

A Curious Result in Geometry

(From David Wells' *Penguin Dictionary of Curious and Interesting Geometry*, page 67)

> Let r be the radius of the incircle of $\triangle ABC$. Let BC be extended an arbitrary distance to K and let another circle of radius r be placed to touch CK and AC as in Figure 1. Draw tangent AD to complete $\triangle ACD$. Continue to construct triangles which radiate from A, each with an incircle of radius r, and having bases that abut along BK.
>
> Then the incircle of every triangle formed by a pair of consecutive triangles in this fan is the same size (for example, the dotted circles in Figure 1).

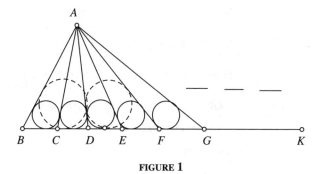

FIGURE 1

It follows similarly that the incircles of the triangles formed by any four consecutive triangles in the fan have the same inradius and, in general, that the inradii of the triangles formed by sets of 2^n consecutive triangles in the fan are all the same. These compound triangles may abut as in Figure 1, they may overlap, or be entirely separated by a gap along BK.

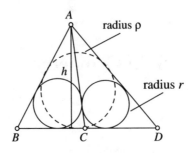

FIGURE 2

Let $\triangle ABD$ be determined by the consecutive triangles ABC and ACD in the fan and suppose the inradius of $\triangle ABD$ is ρ (Figure 2).

Now, the area of a triangle is given by the formula $\Delta = rs$, where r is the inradius and s the semiperimeter of the triangle. Thus

$$\rho = \frac{\Delta}{s},$$

where Δ is the area of $\triangle ABD$.

Denoting the respective areas and semiperimeters of triangles ABC and ACD by Δ_1, Δ_2, s_1, s_2 then

$$\Delta_1 = rs_1, \quad \Delta_2 = rs_2,$$

and

$$\Delta = \Delta_1 + \Delta_2 = r(s_1 + s_2). \tag{1}$$

Now, the sum of the perimeters of the two smaller triangles is

$$2s_1 + 2s_2 = (AB + BC + AC) + (AC + CD + AD)$$
$$= (AB + BD + AD) + 2AC$$
$$= 2s + 2AC.$$

Hence

$$s_1 + s_2 = s + AC,$$

and, recalling (1), we have

$$\Delta = r(s_1 + s_2) = r(s + AC),$$

from which

$$\rho = \frac{\Delta}{s} = \frac{r(s + AC)}{s} = r\left(1 + \frac{AC}{s}\right).$$

Thanks to the formula

$$AC = \sqrt{s(s - BD)},$$

it follows that

$$\frac{AC}{s} = \sqrt{\frac{s - BD}{s}} = \sqrt{1 - \frac{BD}{s}}.$$

Letting the common altitude from A in all the triangles in the fan be h (Figure 2), then

$$\frac{BD}{s} = \frac{\rho \cdot BD}{\rho s} = \frac{\rho \cdot BD}{\Delta} = \frac{\rho \cdot BD}{\frac{1}{2} \cdot BD \cdot h} = \frac{2\rho}{h}$$

giving

$$\frac{AC}{s} = \sqrt{1 - \frac{BD}{s}} = \sqrt{1 - \frac{2\rho}{h}}.$$

Thus

$$\rho = r\left(1 + \frac{AC}{s}\right) = r\left(1 + \sqrt{1 - \frac{2\rho}{h}}\right),$$

which, since h is a constant, gives ρ as a function of just the radius r, showing that ρ is independent of the triangles selected. The conclusion follows.

I found the crucial formula for AC in the amazing collection of Japanese "San Gaku" that was introduced to the Western World in 1989 by Hidetosi Fukagawa (Yokosuka High School, Tokai-City, Japan) and Dan Pedoe (Professor Emeritus, University of Minnesota), published by The Charles Babbage Research Centre, Winnipeg, Canada. It appears there as problem 2.2.5, page 27. Two references are given but no proof. The following proof is due to Professor Dan Velleman.

Referring to Figure 3, let $AB = x$, $BC = y$, $AC = z$, $AD = u$, and $CD = v$. Since triangles ABC and ACD have the same altitude and the same inradius, their areas are proportional to their bases and we have

$$\frac{v}{y} = \frac{\Delta_2}{\Delta_1} = \frac{rs_2}{rs_1} = \frac{s_2}{s_1} = \frac{u + v + z}{x + y + z},$$

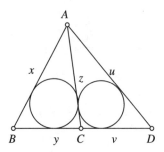

FIGURE 3

which easily reduces to

$$v(x + z) = y(u + z). \tag{1}$$

Letting $\angle ACB = \alpha$ and applying the law of cosines to $\triangle ABC$, we obtain

$$x^2 = y^2 + z^2 - 2yz \cos \alpha,$$

from which

$$\cos \alpha = \frac{y^2 + z^2 - x^2}{2yz}.$$

Observing that $\cos \angle ACD = -\cos \alpha$, applying the law of cosines to $\triangle ACD$ yields

$$u^2 = v^2 + z^2 + 2vz \cos \alpha, \quad \text{and} \quad \cos \alpha = \frac{u^2 - v^2 - z^2}{2vz}.$$

Equating the two expressions for $\cos \alpha$ gives

$$\frac{y^2 + z^2 - x^2}{2yz} = \frac{u^2 - v^2 - z^2}{2vz},$$

and

$$y(u^2 - v^2 - z^2) = v(y^2 + z^2 - x^2). \tag{2}$$

Solving equations (1) and (2) for u and v in terms of x, y, and z, gives

$$u = \frac{xy^2 + 3xz^2 + 2zy^2 + 2z^3 - x^3}{(x + z)^2 - y^2},$$

$$v = \frac{y(y^2 + 3z^2 - x^2 + 2xz)}{(x + z)^2 - y^2}.$$

Now, in terms of x, y, z, u, v,

$$s(s - BD) = \frac{x + y + v + u}{2}\left[\frac{x + y + v + u}{2} - (y + v)\right].$$

The right side of this equation simplifies to z^2 upon substituting for u and v, implying the desired

$$z = AC = \sqrt{s(s - BD)}.$$

From *The Book of Prime Number Records*

The major parts of this section are based on two of the many things learned from Paulo Ribenboim's wonderful *Book of Prime Number Records* (Springer-Verlag, second edition, 1989).

1 Cipolla's Pseudoprimes

Fermat's Simple Theorem asserts that, if n is a prime number, then

$$2^n \equiv 2 \pmod{n}.$$

Thus, as n runs through the positive integers $1, 2, 3, \ldots$, whenever n is a prime number, we have $n | 2^n - 2$. This is such a basic property of prime numbers that any *composite* positive integer n that divides $2^n - 2$ is felt to be acting so much like a prime number that it is called a "2-pseudoprime."

2-pseudoprimes don't exactly stick out among the positive integers and the first one, 341, was discovered as recently as 1819. That is to say, 341 is not a prime, being $11 \cdot 31$, but it does divide $2^{341} - 2$. Once put forward, this is easy to establish on the basis of Fermat's theorem:

observing that $11 | 2^{11} - 2$ since 11 is a prime, then

$$2^{341} - 2 = 2(2^{340} - 1) = 2[(2^{10})^{34} - 1^{34}] = 2(2^{10} - 1)(\ldots)$$
$$= (2^{11} - 2)(\ldots),$$

showing that $11 | 2^{341} - 2$; similarly,

$$2^{341} - 2 = 2[(2^{30})^8 - 1)(\ldots) = 2(2^{30} - 1)(\ldots)$$
$$= (2^{31} - 2)(\ldots), \quad \text{and} \quad 31 | 2^{341} - 2;$$

since 11 and 31 are primes, then $11 \cdot 31 = 341 | 2^{341} - 2$.

For a long time all the known 2-pseudoprimes were odd, and if n is odd and $n|2^n - 2$, it follows that $n|2^{n-1} - 1$. Thus, a 2-pseudoprime came to be thought of as an odd composite positive integer n such that $n|2^{n-1} - 1$.

It is not difficult to show that, if n is an odd 2-pseudoprime, then so is $2^n - 1$ (see the solution to exercise 6 in chapter 1 of my *Mathematical Gems I*, MAA Dolciani Series, vol. 1, 1973). Hence there exist an infinite number of odd 2-pseudoprimes.

It was only in 1950 that D. H. Lehmer discovered that $n|2^n - 2$ for the even integer $n = 161038$. By this time, the term pseudoprime was so firmly attached to odd integers that, even in the light of N. G. W. H. Beeger's 1951 discovery that there also exist an infinite number of even integers n such that $n|2^n - 2$, the unmodified term 2-pseudoprime (or just pseudoprime by itself) is still often used to mean an odd 2-pseudoprime, and a request for clarification is a wise precaution.

Since odd 2-pseudoprimes held the stage for so long, their divisibility criterion came to be adopted for the larger bases. Thus, for a base $a > 2$, an "a-pseudoprime" is a composite positive integer n, odd or even, such that

$$n|a^{n-1} - 1.$$

For even values of $a > 2$, then, an a-pseudoprime is necessarily odd.

We note in passing that composite positive integers n have also been found that divide $a^n - a$ for all integers a. That is to say the same n divides all the numbers $2^n - 2, 3^n - 3, 4^n - 4, \ldots, (-2)^n - (-2), (-3)^n - (-3), \ldots$. These are called *absolute pseudoprimes* and are also known as *Carmichael numbers*. It was proved in 1994 that there exists an infinity of absolute pseudoprimes, the smallest being 561.

Cipolla's Odd a-Pseudoprimes

Finding an odd a-pseudoprime is not a trivial exercise. For example, how how would you go about finding an odd positive integer $n > 1$ that divides $3^{n-1} - 1$? Therefore I feel that I really have something to tell you in presenting Michele Cipolla's wonderfully simple method of constructing an odd a-pseudoprime of any base $a \geq 2$.

Let a base $a \geq 2$ be specified, and let k be any odd prime number that does not divide $a - 1$ or $a + 1$. Then, letting

$$n_1 = \frac{a^k - 1}{a - 1} \quad \text{and} \quad n_2 = \frac{a^k + 1}{a + 1},$$

we already have an odd a-pseudoprime in the product $n = n_1 n_2$.

Clearly n is composite. Observe that

$$n_1 = a^{k-1} + a^{k-2} + \cdots + a + 1,$$

containing an odd number of terms, namely k. Then, since the final term is 1, whether a is odd or even, n_1 can't avoid being odd. Similarly,

$$n_2 = a^{k-1} - a^{k-2} + a^{k-3} - a^{k-4} + - \cdots + a^2 - a + 1,$$

and n_2 is odd. Hence $n = n_1 n_2$ is odd.

Since k is a prime number, Fermat's theorem asserts that $a^k \equiv a \pmod{k}$. From the definition of n_1, then, we have

$$n_1(a - 1) = a^k - 1 \equiv a - 1 \pmod{k},$$

giving

$$(n_1 - 1)(a - 1) \equiv 0 \pmod{k}.$$

Having chosen a prime k that does not divide $a - 1$, it follows that k divides $n_1 - 1$. But, since n_1 is odd, $n_1 - 1$ is even, and so $2k$ divides $n_1 - 1$, that is,

$$n_1 \equiv 1 \pmod{2k}.$$

Similarly,

$$n_2 \equiv 1 \pmod{2k},$$

and we have

$$n = n_1 n_2 \equiv 1 \pmod{2k}.$$

That is to say, $2k$ divides $n - 1$, and therefore, for some positive integer t, we have

$$n - 1 = 2kt.$$

Now, from the definitions of n_1 and n_2, we obtain

$$n = \frac{a^{2k} - 1}{a^2 - 1}, \quad \text{giving} \quad n(a^2 - 1) = a^{2k} - 1,$$

which implies that n divides $a^{2k} - 1$, that is,

$$a^{2k} \equiv 1 \pmod{n}.$$

Finally, then, we have $\pmod{n}$ that

$$a^{n-1} = a^{2kt} = (a^{2k})^t \equiv 1^t \equiv 1,$$

that is $n \mid a^{n-1} - 1$ and n is indeed an odd a-pseudoprime.

An Example Let $a = 3$. Then k may be taken to be any prime that doesn't divide 2 or 4, so let's also take $k = 3$. Then a 3-pseudoprime is given by

$$n = \frac{3^3 - 1}{3 - 1} \cdot \frac{3^3 + 1}{3 + 1} = \frac{26}{2} \cdot \frac{28}{4} = 13 \cdot 7 = 91.$$

In fact, 91 is the smallest 3-pseudoprime.

It is easy to verify that 91 divides $3^{90} - 1$:

$$3^{90} - 1 = (3^6)^{15} - 1 = 729^{15} - 1,$$

and since $729 = 8 \cdot 91 + 1$, then $729 \equiv 1 \pmod{91}$, and therefore

$$3^{90} - 1 \equiv 1^{15} - 1 \equiv 0 \pmod{91}.$$

2 Two Characterizations of Twin Primes

(a) A pair of prime numbers that differ by two, like 17 and 19, is called a pair of twin primes. The first few are

$$(3, 5), (5, 7), (11, 13), (17, 19), (29, 31).$$

It is not known whether there is an infinite number of twin primes. However, it is known that the sum of their reciprocals,

$$\sum \left(\frac{1}{p} + \frac{1}{p + 2} \right),$$

is convergent. Therefore, whether or not they peter out beyond some point, at least they are reasonably rare. Thus it is of some interest that such elusive phenomena can be characterized by a simple elementary condition.

(b) Clement's Characterization   In 1949, P.A. Clement showed that,

> for an odd integer $n \geq 3$, $(n, n + 2)$ is a pair of twin primes if and only if $4[(n - 1)! + 1] + n$ is divisible by $n(n + 2)$.

(i) Sufficiency: Suppose an odd integer $n \geq 3$ satisfies

$$4[(n - 1)! + 1] + n \equiv 0 \pmod{n(n + 2)}. \tag{1}$$

Practically all we need to do is to recall Wilson's Theorem:

> n *is a prime number if and only if* n *divides* $(n - 1)! + 1$.

Since n divides the left side of (1), n divides $4[(n - 1)! + 1]$. Since n is odd, then n divides $(n - 1)! + 1$, and it follows that n is a prime by Wilson's theorem.

It follows similarly that the odd integer $n + 2$ is also a prime. From (1), we have that

$$4[(n-1)! + 1] + n \equiv 0 (\mathrm{mod}\, n + 2),$$

that is,

$$4(n-1)! + (n+2) + 2 \equiv 0 (\mathrm{mod}\, n + 2),$$

and hence

$$4(n-1)! + 2 \equiv 0 (\mathrm{mod}\, n + 2).$$

Multiplying by $(n+1)n$, we have, $(\mathrm{mod}\, n + 2)$, that

$$
\begin{aligned}
0 &\equiv 4(n+1)! + 2n(n+1) \\
&= 4[(n+1)! + 1] + 2n^2 + 2n - 4 \\
&= 4[(n+1)! + 1] + 2(n+2)(n-1) \\
&\equiv 4[(n+1)! + 1] \\
&\equiv [(n+1)! + 1] \quad \text{since } n + 2 \text{ is odd.}
\end{aligned}
$$

Hence $n + 2$ is a prime by Wilson's theorem, and $(n, n + 2)$ is indeed a pair of twin primes.

(ii) Necessity: Suppose $(n, n + 2)$ is a pair of twin primes.

Since $n + 2$ is a prime, we have by Wilson's theorem that

$$(n+1)! + 1 \equiv 0 (\mathrm{mod}\, n + 2).$$

Noting that $(n + 1)n = (n + 2)(n - 1) + 2$, then, substituting for the top two factors in $(n + 1)!$, we obtain $(\mathrm{mod}\, n + 2)$, that

$$
\begin{aligned}
0 &\equiv (n+1)! + 1 \\
&\equiv [(n+2)(n-1) + 2](n-1)! + 1, \\
&= (n+2)(n-1)(n-1)! + 2(n-1)! + 1 \\
&\equiv 2(n-1)! + 1.
\end{aligned}
$$

Thus, for some integer k, we have

$$2(n-1)! + 1 = k(n+2). \tag{2}$$

Now, although k has arisen in connection with $n + 2$, we can get a line on k from the fact that n is a prime: Wilson's theorem gives

$$(n-1)! \equiv -1 (\mathrm{mod}\, n),$$

and so by (2) we get $(\bmod n)$ that

$$k(n+2) = 2(n-1)! + 1 \equiv 2(-1) + 1 \equiv -1,$$
$$kn + 2k \equiv -1, \quad \text{and} \quad 2k + 1 \equiv 0 (\bmod n).$$

Hence, for some integer t, $2k + 1 = tn$.

Finally, by doubling (2), we obtain

$$4(n-1)! + 2 = 2k(n+2),$$

and therefore

$$4[(n-1)! + 1] - 2 = 2k(n+2)$$

and, adding $n + 2$ to each side, we get

$$4[(n-1)! + 1] + n = (2k+1)(n+2)$$
$$= tn(n+2),$$

giving the desired

$$4[(n-1)! + 1] + n \equiv 0 (\bmod n(n+2)).$$

(c) The Characterization of Sergusov, Leavitt, and Mullin

Discoveries by Sergusov in 1971, and by Leavitt and Mullin in 1981, led to the intriguing characterization that

n is the *product* of a pair of twin primes if and only if

$$\varphi(n)\sigma(n) = (n-3)(n+1),$$

where $\varphi(n)$ is Euler's φ-function, which counts the positive integers not exceeding n which are relatively prime to n, and $\varphi(n)$ is the sum of the positive divisors of n, including 1 and n.

Suppose the prime decomposition of n is $p_1^{a_1} p_2^{a_2} \ldots p_k^{a_k}$, where $p_1 < p_2 < \cdots < p_k$. Then

$$\varphi(n) = \prod_{i=1}^{k} p_i^{a_i-1}(p_i - 1) \quad \text{and} \quad \sigma(n) = \prod_{i=1}^{k} \frac{p_i^{a_i+1} - 1}{p_i - 1}.$$

Since the easy derivations of these formulas are given in every elementary textbook, let us not interrupt our story to prove them here.

(a) Necessity: If $n = p(p + 2)$, the product of twin primes, then

$$\varphi(n) = (p - 1)(p + 1) \quad \text{and} \quad \varphi(n) = (p + 1)(p + 3),$$

making

$$\varphi(n)\sigma(n) = (p - 1)(p + 1)^2(p + 3);$$

also

$$(n - 3)(n + 1) = (p^2 + 2p - 3)(p^2 + 2p + 1)$$
$$= (p + 3)(p - 1)(p + 1)^2 = \varphi(n)\sigma(n),$$

as desired.

(b) Sufficiency: Observing that

$$\varphi(n)\sigma(n) = \prod_{i=1}^{k} p_i^{a_i-1}(p_i^{a_i+1} - 1),$$

it remains to show that n is the product of twin primes if

$$\varphi(n)\sigma(n) = \prod_{i=1}^{k} p_i^{a_i-1}(p_i^{a_i+1} - 1) = (n - 3)(n + 1) = n^2 - 2n - 3,$$

which can be written in the form

$$2n + 3 = n^2 - \prod_{i=1}^{k}(p_i^{2a_i} - p_i^{a_i-1}). \tag{3}$$

(i) Clearly each prime p_i divides n and it would also divide the factor $(p_i^{2a_i} - p_i^{a_i-1})$ if a_i were to exceed 1. For $a_i > 1$, then, p_i is a factor of the right side of (3) and therefore it would also have to divide the left side. However, p_i divides n, and therefore the only p_i that can divide $2n + 3$ is $p_i = 3$. It follows that, if $a_i > 1$ then $p_i = 3$, and that a_i must be unity for every prime $p_i \neq 3$.

(ii) Now consider the possibility of $k = 1$. In this case $n = p_1^{a_1}$, and (3) yields

$$2p_1^{a_1} + 3 = p_1^{2a_1} - \left(p_1^{2a_1} - p_1^{a_1-1}\right) = p_1^{a_1-1},$$

where the left side clearly exceeds the right side. Hence $k \geq 2$.

(iii) Next consider the possibility that n is even, i.e., that $p_1 = 2$. Since $k \geq 2$, n has at least a second prime divisor, p_2. Then p_2 must be an odd prime, and that makes the factor $(p_2^{2a_2} - p_2^{a_2-1})$ on the right side of (3) an even number, being the difference between two odd numbers; this is true even if a_2 is as small as 1. Therefore the right side of (3) is an even number and cannot equal the odd number on the left side. Hence $p_1 \geq 3$.

(iv) If $p_1 = 3$, it is easy to see that a_1 is either 1 or 2: To the contrary, suppose $a_1 \geq 3$. In this case, $a_1 - 1 \geq 2$, and the factor $(3^{2a_1} - 3^{a_1-1})$ on the right side of (3) would be divisible by 3^2. Since 3^2 also divides n^2, then 3^2 would be a factor of the right side. However, since $a_1 \geq 3$, then 3^2 divides the first term on the left side, $2n$, but not the second term, 3. Hence 3^2 does not divide the left side, and it follows that a_1 cannot exceed 2.

(v) Summarizing, then, if $p_1 = 3$, we have $p_1^{a_1}$ is either 3 or 3^2, and, recalling that $a_i = 1$ for all $p_i \neq 3$, it follows that either

$$n = 3p_2p_3 \ldots p_k \quad \text{or} \quad n = 3^2 p_2 p_3 \ldots p_k.$$

Otherwise no prime $p_i = 3$, and $n = p_1 p_2 \ldots p_k$. In all cases, then either

(a) $n = p_1 p_2 \ldots p_k$ (where $p_1 \geq 3$), or

(b) $n = 3^2 p_2 \ldots p_k$.

(vi) We have seen that $k \geq 2$. Now let's prove that $k = 2$. Suppose, to the contrary, that $k \geq 3$. In case (a), then, we have the right side of (3)

$$= n^2 - \prod_{i=1}^{k} (p_i^{2a_i} - p_i^{a_i-1})$$

$$= p_1^2 p_2^2 \ldots p_k^2 - (p_1^2 - 1)(p_2^2 - 1) \ldots (p_k^2 - 1)$$

$$> p_1^2 p_2^2 \ldots p_k^2 - (p_1^2 - 1) p_2^2 p_3^2 \ldots p_k^2$$

$$= p_2^2 \ldots p_k^2$$

$$= \frac{p_2 p_3 \ldots p_k}{p_1} \cdot p_1 p_2 \ldots p_k$$

$$> p_3 p_4 \ldots p_k \cdot n \quad \text{(since } p_2 > p_1\text{)}$$

$$\geq 7n \quad \text{(since } p_3 \geq 7 \text{ in view of } p_1 \geq 3\text{)}$$

$$> 2n + 3$$

$$= \text{left side of (3)},$$

a contradiction.

The case of $n = 3^2 p_2 \ldots p_k$ is similar and we conclude that $k = 2$. Thus either

$$n = p_1 p_2, \quad \text{where } p_1 \geq 3, \quad \text{or } n = 3^2 p_2.$$

(vii) Now let's eliminate the case of $n = 3^2 p_2$. In this case, (3) is

$$2(3^2 p_2) + 3 = 81 p_2^2 - (81 - 3)(p_2^2 - 1)$$

i.e.,

$$18 p_2 + 3 = 3 p_2^2 + 78,$$
$$0 = p_2^2 - 6 p_2 + 25,$$

an equation with no real root (the discriminant is -64). Therefore, if condition (3) is satisifed, n can only be the product of two primes $p_1 p_2$.

(viii) Finally, then, in this case, (3) gives

$$2 p_1 p_2 + 3 = p_1^2 p_2^2 - (p_1^2 - 1)(p_2^2 - 1)$$
$$= p_1^2 + p_2^2 - 1,$$

giving

$$4 = p_1^2 - 2 p_1 p_2 + p_2^2 = (p_1 - p_2)^2,$$

revealing that p_1 and p_2 differ by 2, and implying that (p_1, p_2) is indeed a pair of twin primes.

Unfortunately neither of our characterizations is a help in finding twin primes.

Final Comments Sergusov actually gave the two striking results that n is the product of twin primes if and only if

$$\textit{either} \quad \sigma(n) = n + 1 + 2\sqrt{n+1} \quad \textit{or} \quad \varphi(n) = n + 1 - 2\sqrt{n+1}.$$

It seems very likely that Leavitt and Mullin got the idea for their characterization by multiplying these two conditions. In any case, however they might have used Sergusov's brilliant work, Leavitt and Mullin went on to establish the marvelous general result that n is a product of two primes that differ by m if and only if

$$\varphi(n)\sigma(n) = (n-1)^2 - m^2.$$

References

M. Cipolla: Sui numeri composti P, che verificano la congruenza di Fermat $a^{p-1} - 1$; *Annali di Matematica*, (3), 9, 1904.

Sergusov, I.S.A.: On the problem of prime-twins (in Russian), *Jaroslav. Gos. Ped. Inst. Ucen. Zap.*, 82, 1971, 85–86.

Leavitt, W.G. and Mullin, A.A.: Primes Differing by a Fixed Integer, *Mathematics of Computation*, 37, 1981, 581–585.

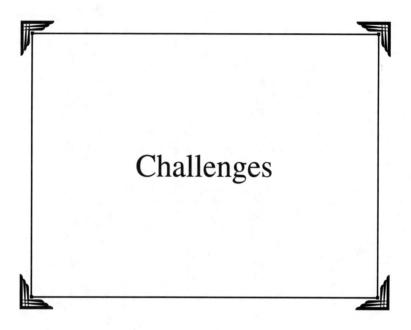

Challenges

Challenges

1. (*Quantum*: Brainteaser B128, Nov.–Dec. 1994)

   Each weight in a set S is ≤ 10 units. S is so constituted that however its weights might be divided into two groups A and B, the weight of at least one of A, B is not bigger than 10. What is the greatest possible total weight of S?

2. (*Quantum*: Brainteaser B129, Nov.–Dec. 1994)

   A 5×9 rectangle R is partitioned into a set S of 10 rectangles with integral dimensions. Prove that some two members of S are congruent.

3. (*Quantum*: Challenges in Physics and Math M168, Mar.–Apr. 1996, proposed by D. Fomin)

   The lengths x, y, z of the sides of a triangle are integers. If two of its altitudes add up to the third altitude, prove that $x^2 + y^2 + z^2$ is a perfect square.

4. (From the 1998 U.S.A. Olympiad)

   The integers $\{1, 2, 3, \ldots, 1998\}$ are put into 999 pairs (a_i, b_i) so that, for each pair, $|a_i - b_i|$ equals either 6 or 1. Prove that the sum

   $$S = |a_1 - b_1| + |a_2 - b_2| + \cdots + |a_{999} - b_{999}|$$

   ends in a 9.

5. (*Mathematics Magazine*, June 2000, Problem 1600, page 239)

   P is a point on an ellipse E. C_1 and C_2 are the circles that pass through P and have centers, respectively, at the foci f_1 and f_2 of E

(Figure 1). Prove that the tangent PQ to E bisects one of the angles between the tangents at P to C_1 and C_2.

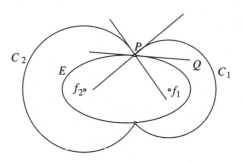

FIGURE 1

6. (*Mathematics Magazine*, June 2000, Quickie Q902, page 241)

Triangle ABC is not obtuse and has sides of integer length satisfying $a < b < c$ (Figure 2). Let altitude BD divide AC into parts of length x and y, as shown. Prove that

$$x - y = 4 \quad \text{if and only if} \quad c - a = 2.$$

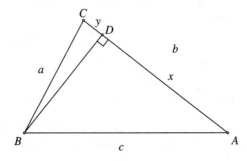

FIGURE 2

7. Prove the intuitively obvious property that the sides of a rectangle inscribed in a non-circular ellipse are respectively parallel to the axes of the ellipse.

8. Given a parabola, just the curve, determine a straightedge and compass construction for its focus.

9. (*Quantum*: From the article "Surprises in Conversion" by I. Kushnir, Mar.–
   Apr., 1996)

   It is obvious that the two segments which join the midpoints of opposite sides of a parallelogram are respectively equal in length to the sides
   of the parallelogram, and so their sum is one-half the perimeter of the
   parallelogram (Figure 3). Prove the converse theorem:

   If K, L, M, N are the midpoints of the sides of quadrilateral $ABCD$
   and $KM + LN$ is equal to one-half the perimeter of $ABCD$, prove that
   $ABCD$ must be a parallelogram.

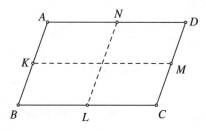

FIGURE 3

10. (*Quantum*: From the article "Surprises in Conversion" by I. Kushnir, Mar.–
    Apr., 1996)

    If P is the orthocenter of acute-angled $\triangle ABC$, then, in cyclic quadri-
    lateral $AEFC$, chord EF subtends equal angles x at A and C (Figure 4(a));
    similarly, in cyclic quadrilateral $ABFG$, chord FG subtends equal angles y
    at A and B.

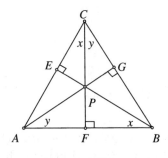

FIGURE 4(a)

Prove the converse theorem:

If P is a point inside an acute-angled triangle ABC such, in Figure 4(b), that the two angles marked x are equal and the two angles marked y are equal, prove that P must be the orthocenter.

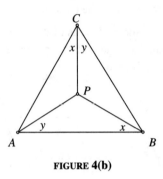

FIGURE 4(b)

11. Eddie's Problem (see part 4 of §14, From the Desk of Liong-shin Hahn)

Given a 5×5 set of lattice points (Figure 5), find a set of five circles which pass through each of the 25 points at least once.

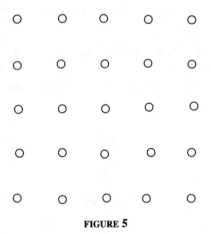

FIGURE 5

12. (From the outstanding collection *Mathematical Olympiad Problems* by Titu Andreescu and Răzvan Gelca (Birkhäuser Boston, 2000) page 19, #6, which is part of Problem 1 of the Morning Session of the 1963 Putnam Competition)

$P_1, P_2, \ldots, P_{12}$ are the successive vertices of a regular dodecagon. Prove that $P_1 P_9$, $P_2 P_{11}$, and $P_4 P_{12}$ are concurrent (Figure 6).

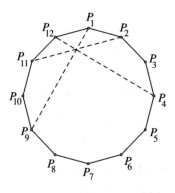

FIGURE 6

13. (From *Mathematical Olympiad Problems* by Titu Andreescu and Răzvan Gelca (Birkhäuser Boston, 2000) page 9, #10, from the 1998 Asian-Pacific Mathematical Olympiad)

In Figure 7, AD is an altitude in $\triangle ABC$. E and F are points on a line DT such that AE is perpendicular to BE and AF is perpendicular to CF. M is the midpoint of BC and N is the midpoint of EF. Prove that AN is perpendicular to MN.

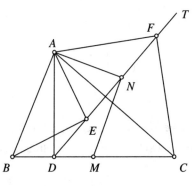

FIGURE 7

14. (From the 2001 Putnam Competition)

Consider a set S and a binary operation $\ominus$ on S (that is, for each a, b in S, $a \ominus b$ is in S). Given that

$$(a \ominus b) \ominus a = b \quad \text{for all } a, b \text{ in } S,$$

prove that

$$a \ominus (b \ominus a) = b \quad \text{for all } a, b \text{ in } S.$$

15. (An old chestnut)

Determine the value of the infinite continued fraction

$$y = 1 + \cfrac{1}{2 + \cfrac{1}{2 + \cfrac{1}{2 + \cdots}}}.$$

16. (Problem 721, *College Mathematics Journal*, March 2002, page 150)

I is the incenter of $\triangle ABC$ and D is the point of contact of the incircle on side AB (Figure 8). ID is extended to H so that DH equals the semiperimeter of the triangle. Prove that $AIBH$ is cyclic if and only if angle C is a right angle.

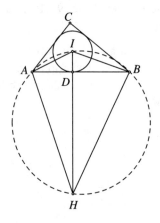

FIGURE 8

17. (The upside-down theorem of Pythagoras, *CMJ*, March 2002, page 170)

In right triangle ABC, $CD = x$ is the altitude to the hypotenuse AB (Figure 9). Prove that

$$a^{-2} + b^{-2} = x^{-2}.$$

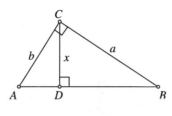

FIGURE 9

18. (Problem 955, *Pi Mu Epsilon Journal*, Spring 2000, page 105)

    Let

    $$S = a + ar + ar^2 + \cdots + ar^{n-1},$$

    where a, r and n are positive integers, and r and n are greater than 1. If S is a prime number, prove that n must also be a prime number.

19. (Problem 988, *Pi Mu Epsilon Journal*, Spring 2001, page 223)

    For what positive integers n is

    $$1^3 - 2^3 + 3^3 - 4^3 + - \cdots + (-1)^{n+1} n^3$$

    a perfect square?

20. (Problem 990, *Pi Mu Epsilon Journal*, Spring 2001)

    Identify all triangles ABC such that $\cos^2 A + \cos^2 B + \cos^2 C = 1$.

21. (*Quantum*: A problem of N. Vasilyev, Problem M309 of the Challenges, Nov./Dec. 2000, page 13)

    Given three points A, B, C in the plane, construct a straight line L through C such that the product xy of the perpendiculars x and y to L from A and B is a maximum.

22. (*Quantum*: A problem of N. Vasilyev, Problem M314, Jan./Feb. 2001, page 33)

    If the product of two positive numbers is greater than their sum, prove that their sum must exceed 4.

23. (*Quantum*: A problem from "Where Do Problems Come From?" by I. Sharygin, Jan./Feb. 2001, pages 18–28)

    In $\triangle ABC$, angle B is 120° (Figure 10). BM is the bisector of angle B and the bisector CK of exterior angle BCD meets AB extended at K. KM crosses BC at P. How big is $\angle APM$?

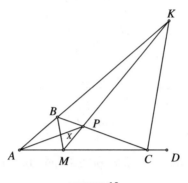

FIGURE 10

24. (*Quantum*: From "Problems Teach Us How To Think" by V. Proizvolov, Jan./Feb. 2002, pages 42–45)

ABCDE is a zigzag line in a circle such that the angles at B, C, and D are each 45° (Figure 11). Prove that

$$AB^2 + CD^2 = BC^2 + DE^2.$$

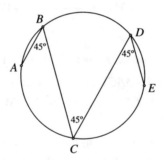

FIGURE 11

25. (*Quantum*: A problem from the article "Taking on Triangles" by A. Kanel and A. Kovaldzhi, Mar./Apr. 2001, pages 10–16)

Ali Baba arrived at a cave where there are gold, diamonds, and a trunk. If the trunk is filled with gold, it weighs 200 kilograms; if it is filled with diamonds, it weights 40 kilograms; the empty trunk is weightless. A kilogram of gold is worth 20 dinars and a kilogram of diamonds is worth 60 dinars. What is the greatest value of the treasure that Ali can take away if he can carry only 100 kilograms?

26. Prove the surprising result that $a^{1/\ln a}$ is a constant for all positive real numbers $a \neq 1$.

27. A Problem of Rex H. Wu, Brooklyn, New York (*The College Mathematics Journal*, March 2003, page 115)

   Use Figure 12 to verify each of the equations

   (a) $\tan^{-1}(1/2) + \tan^{-1}(1/3) = \pi/4$,

   (b) $\tan^{-1}(3) - \tan^{-1}(1/2) = \pi/4$,

   (c) $\tan^{-1}(2) - \tan^{-1}(1/3) = \pi/4$,

   (d) $\tan^{-1}(1) + \tan^{-1}(1/2) + \tan^{-1}(1/3) = \pi/2$,

   (e) $\tan^{-1}(1) + \tan^{-1}(2) + \tan^{-1}(3) = \pi$.

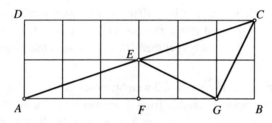

FIGURE 12

Solutions To the Challenges

1. (*Quantum*: Brainteaser B128, Nov.–Dec. 1994)

    Each weight in a set S is ≤ 10 units. S is so constituted that however its weights might be divided into two groups A and B, the weight of at least one of A, B is not bigger than 10. What is the greatest possible total weight of S?

    Clearly S could consist of 3 weights of size 10, implying a maximum ≥ 30. Now, suppose weights from S are put into A until the **first** time its weight goes over 10. Since no weight exceeds 10, the first time this happens cannot take the weight of A to more than 20. Thus, if the total weight of S were to exceed 30, there would remain more than $30 - 20 = 10$ in the complementary set B, making the weights of both A and B in excess of 10, a contradiction. Hence the maximum weight of S is indeed 30.

2. (*Quantum*: Brainteaser B129, Nov.–Dec. 1994)

    A 5×9 rectangle R is partitioned into a set S of 10 rectangles with integral dimensions. Prove that some two members of S are congruent.

    The possible rectangles in S, listed in order of increasing area, are

    $$\{1 \times 1, 1 \times 2, 1 \times 3, 1 \times 4, 2 \times 2, 1 \times 5, 1 \times 6, 2 \times 3, 1 \times 7, 1 \times 8, 2 \times 4, \ldots\},$$

    with areas respectively of $\{1, 2, 3, 4, 4, 5, 6, 6, 7, 8, 8, \ldots\}$.
    Thus the smallest total area of 10 *different* rectangles is

    $$1 + 2 + 3 + 4 + 4 + 5 + 6 + 6 + 7 + 8 = 46 > 5 \cdot 9,$$

    exceeding the area of a 5 by 9 rectangle. Hence a duplication is unavoidable.

3. (*Quantum*: Challenges in Physics and Math M168, Mar.–Apr. 1996, proposed with a similar solution by D. Fomin)

The lengths x, y, z of the sides of a triangle are integers. If two of its altitudes add up to the third altitude, prove that $x^2 + y^2 + z^2$ is a perfect square.

Let the lengths of the altitudes be a, b, c, where $a + b = c$, and let the area of the triangle be $\Delta = \frac{1}{2}ax = \frac{1}{2}by = \frac{1}{2}cz$. Hence

$$a = \frac{2\Delta}{x}, \quad b = \frac{2\Delta}{y}, \quad c = \frac{2\Delta}{z},$$

and $a + b = c$ gives

$$\frac{2\Delta}{x} + \frac{2\Delta}{y} = \frac{2\Delta}{z},$$
$$\frac{1}{x} + \frac{1}{y} = \frac{1}{z},$$
$$yz + xz = xy,$$

and

$$xy - yz - zx = 0.$$

Then

$$x^2 + y^2 + z^2 = x^2 + y^2 + z^2 + 2(xy - yz - zx)$$
$$= (x + y - z)^2,$$

a perfect square.

4. (From the 1998 U.S.A. Olympiad)

The integers $\{1, 2, 3, \ldots, 1998\}$ are put into 999 pairs (a_i, b_i) so that, for each pair, $|a_i - b_i|$ equals either 6 or 1. Prove that the sum

$$S = |a_1 - b_1| + |a_2 - b_2| + \cdots + |a_{999} - b_{999}|$$

ends in a 9.

Let n be the number of times $|a_i - b_i| = 6$; then $999 - n$ is the number of times $|a_i - b_i| = 1$. Therefore

$$S = 6 \cdot n + 1 \cdot (999 - n) = 5n + 999,$$

which is a number that ends in a 9 if and only if n is even.

There are 999 even numbers and 999 odd numbers among the a_i and b_i and it takes two of the same parity to make $|a_i - b_i| = 6$ but only one of each parity to make $|a_i - b_i| = 1$. Thus the cases $|a_i - b_i| = 6$

consume some even number of the 999 even numbers available, leaving an odd number of them for the cases of $|a_i - b_i| = 1$. Since these latter cases use them up one at a time, the number of such cases, $999 - n$, must be odd, and this means that n has to be even, as desired.

5. *Mathematics Magazine*, June 2000, Problem 1600, page 239)

P is a point on an ellipse E. C_1 and C_2 are the circles that pass through P and have centers, respectively, at the foci f_1 and f_2 of E (Figure 1). Prove that the tangent PQ to E bisects one of the angles between the tangents at P to C_1 and C_2.

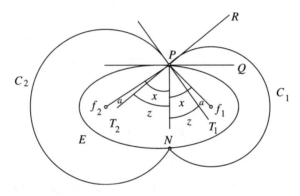

FIGURE 1

Let the tangents be PQ, PT_1, and PT_2, as shown in Figure 1, and let PN be the normal to the ellipse at P.

By the reflector property of the ellipse, PN bisects the angle between the focal radii, giving equal angles x, as shown. Now, tangent T_2P is perpendicular to radius Pf_1, and tangent T_1P is perpendicular to radius Pf_2. Subtracting these right angles from angle f_2Pf_1 makes the small angles marked a equal. Subtracting these as from the xs, then, gives equal angles z. That is to say, the normal PN bisects one of the angles between the tangents to the circles, and therefore the tangent PQ, being perpendicular to the normal, bisects the other angle between them.

6. (*Mathematics Magazine*, June 2000, Quickie Q902, page 241)

Triangle ABC is not obtuse and has sides of integer length satisfying $a < b < c$ (Figure 2). Let altitude BD divide AC into parts of length x and y, as shown. Prove that

$$x - y = 4 \quad \text{if and only if} \quad c - a = 2.$$

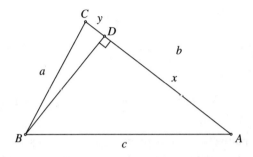

FIGURE 2

By the theorem of Pythagoras,

$$BD^2 = c^2 - x^2 = a^2 - y^2,$$

and we have

$$c^2 - a^2 = x^2 - y^2,$$
$$(c - a)(c + a) = (x - y)(x + y) = (x - y) \cdot b. \tag{4}$$

(i) Suppose $c - a = 2$, that is, $c = a + 2$.

Since a, b, c are integers and $a < b < c$, this makes $b = a + 1$. Thus

$$c = a + 2 \quad \text{and} \quad c + a = 2a + 2 = 2(a + 1) = 2b.$$

Hence (4) yields

$$2 \cdot 2b = (x - y)b, \quad \text{giving} \quad x - y = 4.$$

(ii) Suppose $x - y = 4$:

Then (4) gives

$$(c - a)(c + a) = 4b. \tag{5}$$

Now, the triangle inequality asserts that $c + a > b$, and therefore (5) yields $c - a < 4$, limiting its values to 1, 2, or 3. We need to eliminate cases 1 and 3. Clearly $c - a = 1$ doesn't leave room for b in between a and c. Therefore, suppose $c - a = 3$. In this case, (5) gives

$$c + a = \tfrac{4}{3}b.$$

Then

$$2c = (c - a) + (c + a) = 3 + \tfrac{4}{3}b$$

and

$$6c = 9 + 4b,$$

which is impossible because the left side is even and the right side is odd.

7. Prove the intuitively obvious property that the sides of a rectangle inscribed in a non-circular ellipse are respectively parallel to the axes of the ellipse.

The reason the axes of an ellipse are called axes is because they are "axes of symmetry," and moreover, they are the only axes of symmetry the ellipse possesses. The inscribed rectangle also has two axes of symmetry, namely its two "midlines" that join the midpoints of opposite sides and which are clearly respectively parallel to the sides of the rectangle, and we would like to show that the same two lines are the axes of both the ellipse and the rectangle.

Referring to Figure 3, let $ABCD$ be the inscribed rectangle. Let M and N be the midpoints of sides AB and CD, and let MN extended give chord EF of the ellipse. Then EF bisects the parallel chords AB and CD and is therefore the "diameter" of the set of chords which are parallel to AB and CD. (Recall that a *diameter* of a conic is the locus of the midpoints of all its chords in a given direction. It turns out that all such midpoints lie on a straight line.) Now, all diameters of an ellipse go through the center of the ellipse and are bisected there. Hence the midpoint O of EF is the center of the ellipse, and hence every chord through O is bisected there.

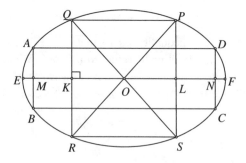

FIGURE 3

Recall that our goal is to show that *EF* is an axis of symmetry of the ellipse. Accordingly, let *P* be an arbitrary point on the ellipse. Let *PO* meet the ellipse at *R* and let the chord through *R* perpendicular to *EF* meet *EF* at *K* and the ellipse at *Q*. Let *QO* meet the ellipse at *S*, and let *PS* cross *EF* at *L*.

Since *RQ* is perpendicular to *EF*, as is *AB*, then *RQ* is parallel to *AB* and is therefore another chord that is bisected by diameter *EF*. With *QK = KR*, triangles *QKO* and *RKO* are congruent (SAS), giving *QO = RO*. But the center *O* bisects chords *PR* and *QS*, and so all four segments *OP, OQ, OR, OS* are equal.

It follows easily that *PS* is parallel to *QR*, and therefore also to *AB*, in which case diameter *EF* bisects it, making *PL = LS*. Furthermore, being parallel to *AB*, *PS* is perpendicular to *EF*, and hence *P* reflects in *EF* to the point *S* on the ellipse. Thus *EF* is indeed an axis of symmetry of the ellipse. Similarly for the other midline, and our argument is complete.

8. Given a parabola, just the curve, determine a straightedge and compass construction for its focus.

A simple construction may be based on the reflector property of a parabola and a couple of friendly properties of diameters. Recall that a *diameter* of a conic is the locus of the midpoints of all its chords in a given direction. It turns out that all such midpoints lie on a straight line and that, for a parabola,

1. Every diameter is parallel to the axis.

2. The tangent at the end of a diameter is parallel to the set of chords that is bisected by the diameter.

First, draw any two parallel chords *AB* and *CD* (Figure 4). Determine their midpoints *L* and *M*, thus obtaining their diameter *LM*. Let *LM* meet the parabola at *K*. Then the line *SKT* which is parallel to *AB* is the tangent at *K* and the perpendicular *KN* to *ST* is therefore the normal at *K*.

Now, diameter *LM* is parallel to the axis and, by the reflector property, a line parallel to the axis is reflected through the focus. Hence, duplicating ∠*MKN* as ∠*NKP* gives a line *KP* that goes through the focus. Repeating with a different pair of parallel chords completes the construction with a second line through the focus.

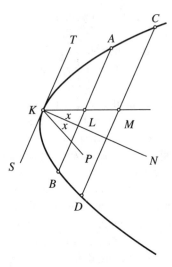

9. (*Quantum*: From the article "Surprises in Conversion" by I. Kushnir, Mar.–Apr. 1996)

It is obvious that the two segments which join the midpoints of opposite sides of a parallelogram are respectively equal in length to the sides of the parallelogram, and so their sum is one-half the perimeter of the parallelogram (Challenge Figure 3). Prove the converse theorem:

If K, L, M, N are the midpoints of the sides of quadrilateral $ABCD$ and $KM + LN$ is equal to one-half the perimeter of $ABCD$, prove that $ABCD$ must be a parallelogram (Figure 5).

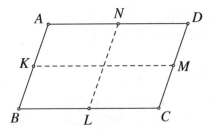

FIGURE 5

If E is the midpoint of the diagonal AC (Figure 6), then KE, which joins the midpoints of two sides of $\triangle ABC$, is parallel to BC, and similarly EM is parallel to AD. Thus, if we can show that E lies on KM, then both AD and BC would be parallel to KM, making them parallel to each other.

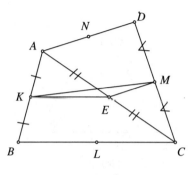

Observe that KE is not only parallel to BC but half as long; similarly EM is half AD. Thus

$$KE + EM = \tfrac{1}{2}(BC + AD).$$

Now, the triangle inequality gives $KE + EM \geq KM$, and so

$$KM \leq \tfrac{1}{2}(BC + AD). \tag{a}$$

Similarly,

$$LN \leq \tfrac{1}{2}(AB + CD). \tag{b}$$

Now, it is given that

$$KM + LN = \tfrac{1}{2} \text{ (the perimeter of } ABCD),$$

which could not be true if either (a) or (b) were a strict inequality. Thus (a) gives

$$KM = \tfrac{1}{2}(BC + AD) = KE + EM,$$

implying that E indeed lies on KM, and BC is parallel to AD. Similarly AB is parallel to CD and $ABCD$ is a parallelogram.

10. (*Quantum*: From the article "Surprises in Conversion" by I. Kushnir, Mar.–Apr., 1996)

    (Figure 7) If P is a point inside triangle ABC such that the two angles marked x are equal and the two angles marked y are equal, prove that P must be the orthocenter.

    Let CP be extended to meet AB at D and the circumcircle of the triangle at E (Figure 8). Then on chord EB we have

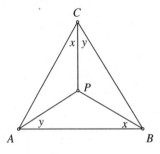

FIGURE 7

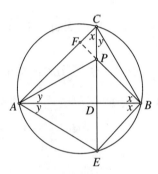

FIGURE 8

$$\angle EAB = \angle ECB = y = \angle BAP;$$

similarly on chord AE,

$$\angle ABE = \angle ACE = x = \angle ABP.$$

Thus, reflecting in AB would carry AE along AP and EB along BP and thus take E to P. That is to say, AB is the perpendicular bisector of PE, implying that CPD is an altitude.

Let BP be extended to F on AC. Since $\angle ADC$ is a right triangle, $\angle CAD + \angle ACD = 90°$, i.e., $\angle CAD + x = 90°$. Hence in $\triangle FAB$ the angles at A and B add to $90°$, making the angle at F a right angle. Hence BPF is also an altitude, making P the orthocenter.

11. Eddie's Problem (See §14, From the Desk of Liong-shin Hahn)

Given a 5×5 array of lattice points, find a set of five circles which pass through each point at least once.

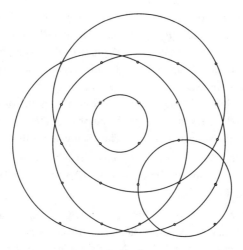

FIGURE 9

12. (From the outstanding collection *Mathematical Olympiad Problems* by
    Titu Andreescu and Răzvan Gelca (Birkhäuser Boston, 2000) page 19,
    #6, which is part of Problem 1 of the Morning Session of the 1963 Putnam
    Competition)

    $P_1, P_2, \ldots, P_{12}$ are the successive vertices of a regular dodecagon.
    Prove that $P_1 P_9$, $P_2 P_{11}$, and $P_4 P_{12}$ are concurrent.

    Being regular, the dodecagon is cyclic. Let O be the center of the
    dodecagon and r its circumradius. Let OP_{11} meet $P_9 P_1$ at S, and let $P_{12} P_4$
    cross OP_2 at T (Figure 10).

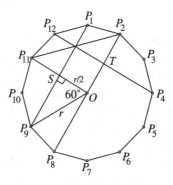

FIGURE 10

Now, each side of the dodecagon subtends $30°$ at O, making $\angle P_9OP_{11} = 60°$ and OP_{11} perpendicular to OP_2. But chord P_9P_1 is parallel to diameter P_8P_2, and so OP_{11} and P_9P_1 are perpendicular. Thus $\triangle P_9OS$ is a 30–60–90 degree triangle with hypotenuse r. Hence $OS = \frac{r}{2}$.

Therefore, in $\triangle OP_{11}P_2$, P_9P_1 goes through the midpoint S of side OP_{11}, is parallel to side OP_2, and hence it bisects the third side $P_{11}P_2$. Similarly, P_4P_{12} goes through the midpoint T of OP_2, is parallel to OP_{11} and therefore it also bisects $P_{11}P_2$. The conclusion follows.

13. (From *Mathematical Olympiad Problems* by Titu Andreescu and Răzvan Gelca (Birkhäuser Boston, 2000) page 9, #10, from the 1998 Asian-Pacific Mathematical Olympiad)

In Figure 11, AD is an altitude in $\triangle ABC$. E and F are points on a line DT such that AE is perpendicular to BE and AF is perpendicular to CF. M is the midpoint of BC and N is the midpoint of EF. Prove that AN is perpendicular to MN.

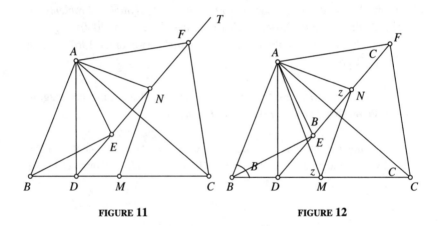

FIGURE 11   FIGURE 12

Since AB subtends a right angle at D and E, $ABDE$ is cyclic, implying the exterior angle at E, $\angle AEF$, is equal to the interior angle B (i.e., $\angle ABD$) at the opposite vertex (Figure 12). Similarly, the right angles at D and F make $ADCF$ cyclic, and chord AD subtends equal angles C at C and F.

Thus triangles ABC and AEF are similar, and hence the medians AM and AN meet the opposite sides EF and BC at the same angles; that is to say, $\angle AMB = \angle ANE \, (= z)$ (the easy proof of this is left as an exercise).

Therefore AD subtends the same angle z at M and N, making $ADMN$ cyclic, and since the angle at D is a right angle, so is the supplementary angle at N.

14. (From the 2001 Putnam Competition)

Consider a set S and a binary operation $\ominus$ on S (that is, for each a, b in S, $a \ominus b$ is in S). Given that

$$(a \ominus b) \ominus a = b \quad \text{for all } a, b \text{ in } S, \tag{1}$$

prove that

$$a \ominus (b \ominus a) = b \quad \text{for all } a, b \text{ in } S.$$

Interchanging a and b in (1), we find $(b \ominus a) \ominus b = a$. Substituting this for the initial a in $a \ominus (b \ominus a)$, we get

$$a \ominus (b \ominus a) = \big[(b \ominus a) \ominus b\big] \ominus (b \ominus a),$$

which is the left side of (1) with $(b \ominus a)$ in place of a. By (1), then, the result is still b, as desired.

15. (An old chestnut)

Determine the value of the infinite continued fraction

$$y = 1 + \cfrac{1}{2 + \cfrac{1}{2 + \frac{1}{2 + \cdots}}}.$$

Let

$$x = 2 + \cfrac{1}{2 + \cfrac{1}{2 + \frac{1}{2 + \cdots}}}.$$

Then

$$y = x - 1 = 1 + \frac{1}{x},$$
$$x^2 - 2x - 1 = 0,$$
$$x = \frac{2 \pm \sqrt{8}}{2} = 1 \pm \sqrt{2}.$$

Since x is positive, then $x = 1 + \sqrt{2}$ and $y = x - 1 = \sqrt{2}$.

16. (Problem 721, *College Mathematics Journal*, March 2002, page 150)

I is the incenter of $\triangle ABC$ and D is the point of contact of the incircle on side AB. ID is extended to H so that DH equals the semiperimeter of the triangle. Prove that $AIBH$ is cyclic if and only if angle C is a right angle.

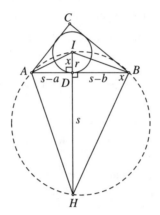

FIGURE 13

As usual, let a, b, c denote the lengths of the sides of $\triangle ABC$ which are respectively opposite vertices A, B, C.

It is well known that $AD = s - a$, $BD = s - b$ (Figure 13), and that the area of the $\triangle ABC = rs = \sqrt{s(s-a)(s-b)(s-c)}$.

$AIBH$ is cyclic if and only if AH subtends the same angle at I and B, that is, if and only if triangles ADI and DHB are similar, that is, if and only if

$$\frac{s-a}{s} = \frac{r}{s-b},$$

$$(s-a)(s-b) = rs = \sqrt{s(s-a)(s-b)(s-c)},$$

$$(s-a)(s-b) = s(s-c),$$

$$s^2 - (a+b)s + ab = s^2 - sc,$$

$$ab = (a+b-c)s = \frac{(a+b-c)(a+b+c)}{2},$$

$$2ab = (a+b)^2 - c^2,$$

$$c^2 = a^2 + b^2,$$

that is, if and only if angle C is a right angle.

17. (The upside-down theorem of Pythagoras, *CMJ*, March 2002, page 170)

In right triangle ABC, $CD = x$ is the altitude to the hypotenuse AB (Figure 14). Prove that

$$a^{-2} + b^{-2} = x^{-2}.$$

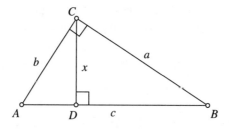

FIGURE 14

Clearly,

$$\triangle ABC = \tfrac{1}{2}cx = \tfrac{1}{2}ab.$$

Doubling, squaring, and noting that $c^2 = a^2 + b^2$, then

$$c^2x^2 = (a^2 + b^2)x^2 = a^2b^2,$$

from which

$$\frac{a^2 + b^2}{a^2b^2} = \frac{1}{x^2},$$

that is,

$$\frac{1}{b^2} + \frac{1}{a^2} = \frac{1}{x^2},$$

as desired.

18. (Problem 955, *Pi Mu Epsilon Journal*, Spring 2000, page 105)

Let

$$S = a + ar + ar^2 + \cdots + ar^{n-1},$$

where a, r and n are positive integers, and r and n are greater than 1. If S is a prime number, prove that n must also be a prime number.

Clearly, a divides S, and since S is a prime, then a is either 1 or S. But clearly $S > a$, and so $a = 1$, and we have

$$S = 1 + r + r^2 + \cdots + r^{n-1} = \frac{r^{n-1}}{r - 1},$$

giving

$$(r - 1)S = r^n - 1.$$

Proceeding indirectly, suppose n is not a prime number. Then for some positive integers a and b, each greater than 1, we have

$$n = ab.$$

In this case,

$$
\begin{aligned}
(r - 1)S = r^{ab} - 1 &= (r^a)^b - 1 \\
&= (r^a - 1)\big[(r^a)^{b-1} + (r^a)^{b-2} + \cdots + 1\big] \\
&= (r - 1)[r^{a-1} + r^{a-2} + \cdots + 1] \\
&\quad \cdot \big[(r^a)^{b-1} + (r^a)^{b-2} + \cdots + 1\big].
\end{aligned}
$$

Since $r > 1$, then $r - 1 \neq 0$ and we obtain

$$S = [r^{a-1} + r^{a-2} + \cdots + 1]\big[(r^a)^{b-1} + (r^a)^{b-2} + \cdots + 1\big],$$

where each large bracket is greater than 1 since it contains at least two terms. Thus S is not a prime number and the conclusion follows by contradiction.

19. (Problem 988, *Pi Mu Epsilon Journal*, Spring 2001, page 223)

For what positive integers n is the sum

$$S_n = 1^3 - 2^3 + 3^3 - 4^3 + - \cdots + (-1)^{n+1} n^3$$

a perfect square?

One can hardly fail to be curious about the answer, but one need not wonder for long for the problem yields to the straightforward approach of adding up the series.

First we observe that each of the differences

$$(1^3 - 2^3), (3^3 - 4^3), \ldots, \big[(n - 1)^3 - n^3\big]$$

is negative. Thus, S_n is negative for n even and consequently not a perfect square. Hence n must be odd.

Recall that the sum of the cubes of the first n positive integers is

$$\left[\frac{n(n + 1)}{2}\right]^2.$$

Thus, adding and subtracting the cubes of the even integers, we have

$$S_n = (1^3 + 2^3 + 3^3 + \cdots + n^3) - 2[2^3 + 4^3 + \cdots + (n-1)^3]$$

$$= \left[\frac{n(n+1)}{2}\right]^2 - 2.8\left[1^3 + 2^3 + 3^3 + \cdots + \left[\frac{n-1}{2}\right]^3\right]$$

$$= \left[\frac{n(n+1)}{2}\right]^2 - 16\left(\frac{\frac{n-1}{2} \cdot \frac{n+1}{2}}{2}\right)^2$$

$$= \left[\frac{n(n+1)}{2}\right]^2 - 4\left(\frac{n-1}{2}\right)^2 \cdot \left(\frac{n+1}{2}\right)^2$$

$$= \left[\frac{n+1}{2}\right]^2 \cdot \left[n^2 - 4\left(\frac{n-1}{2}\right)^2\right]$$

$$= \left[\frac{n+1}{2}\right]^2 [n^2 - (n-1)^2]$$

$$= \left[\frac{n+1}{2}\right]^2 (2n-1).$$

In order to be a square, the odd number $2n - 1$ must be a square, and perforce the square of an odd number:

$$2n - 1 = (2k - 1)^2 = 4k^2 - 4k + 1$$

giving

$$n = 2k^2 - 2k + 1.$$

Thus any positive integer n of the form $2k(k-1) + 1$ makes S_n a perfect square (e.g., $n = 1, 5, 13, 25, \ldots$).

20. (Problem 990, *Pi Mu Epsilon Journal*, Spring 2001)

Identify all triangles ABC such that $\cos^2 A + \cos^2 B + \cos^2 C = 1$.

The key is to establish the triangle identity

$$\cos^2 A + \cos^2 B + \cos^2 C = 1 - 2\cos A \cos B \cos C.$$

To this end, observe that, since $C = 180° - (A + B)$, then $\cos C = -\cos(A + B)$ and hence

$$\cos^2 C = \cos^2(A + B) = (\cos A \cos B - \sin A \sin B)^2,$$

and

$$\cos^2 C = \cos^2 A \cos^2 B - 2\cos A \cos B \sin A \sin B + \sin^2 A \sin^2 B. \quad (1)$$

Now,

$$\sin^2 A \sin^2 B = (1 - \cos^2 A)(1 - \cos^2 B)$$
$$= 1 - \cos^2 A - \cos^2 B + \cos^2 A \cos^2 B.$$

Therefore, substituting this in (1), we get

$$\cos^2 C = \cos^2 A \cos^2 B - 2\cos A \cos B \sin A \sin B$$
$$+ 1 - \cos^2 A - \cos^2 B + \cos^2 A \cos^2 B$$
$$= 1 + 2\cos^2 A \cos^2 B - 2\cos A \cos B \sin A \sin B$$
$$- \cos^2 A - \cos^2 B,$$

and, transposing $\cos^2 A$ and $\cos^2 B$, we obtain

$$\cos^2 A + \cos^2 B + \cos^2 C$$
$$= 1 + 2\cos^2 A \cos^2 B - 2\cos A \cos B \sin A \sin B$$
$$= 1 + 2\cos A \cos B[\cos A \cos B - \sin A \sin B]$$
$$= 1 + 2\cos A \cos B \cos(A + B)$$
$$= 1 - 2\cos A \cos B \cos C.$$

As a result, $\cos^2 A + \cos^2 B + \cos^2 C = 1$ if and only if one of $\cos A$, $\cos B$, $\cos C$ vanishes. That is to say, the condition holds if and only if the triangle is right-angled.

21. (*Quantum*: A problem of N. Vasilyev, Problem M309 of the Challenges, Nov./Dec. 2000, page 13)

Given three points A, B, C in the plane, construct a straight line L through C such that the product xy of the perpendiculars x and y to L from A and B is a maximum.

Let $BC = a$ and $AC = b$, and let L lie within the angle C of $\triangle ABC$. Suppose L divides $\angle C$ into parts u and v (Figure 15).
Then

$$x = b \sin u \quad \text{and} \quad y = a \sin v.$$

Hence

$$xy = ab(\sin u)(\sin v),$$
$$= ab\left\{\tfrac{1}{2}\big[\cos(u - v) - \cos(u + v)\big]\right\},$$

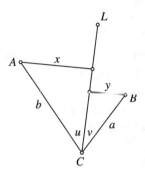

FIGURE 15

where a, b, and $u + v$ are constants. Thus xy is greatest when $\cos(u - v)$ is greatest, namely for $u - v = 0$, i.e., $u = v$. For L lying across triangle ABC, then, xy is greatest when L bisects $\angle C$.

We observe that, in this case,

$$\text{maximum } xy = P_1 = ab \sin^2 \frac{C}{2}.$$

If L lies outside $\triangle ABC$, a similar analysis shows that xy is greatest when L bisects the exterior angle of $\triangle ABC$ at C, and that the maximum value of xy is

$$P_2 = ab \sin^2 \left(\tfrac{1}{2} \text{ the exterior angle at } C \right).$$

Now, if $\angle C$ is acute, then its exterior angle is greater than $\angle C$ and $P_2 > P_1$, implying that only the exterior bisector of $\angle C$ gives the maximum. If $\angle C$ is obtuse, the reverse is true, $P_1 > P_2$, and only the interior bisector of $\angle C$ gives the maximum. And if $\angle C$ is a right angle, then $P_1 = P_2$, and both the interior and exterior bisectors of $\angle C$ provide the maximum.

22. (*Quantum*: A problem of N. Vasilyev, Problem M314, Jan./Feb. 2001, page 33)

If the product of two positive numbers is greater than their sum, prove that their sum must exceed 4.

Let the numbers be x and y. Then $xy > x + y$.
From the arithmetic mean-geometric mean inequality we have

$$\frac{x + y}{2} \geq \sqrt{xy}.$$

Doubling and squaring gives $(x + y)^2 \geq 4xy > 4(x + y)$. Dividing by the positive number $x + y$, we obtain the desired $x + y > 4$.

23. (*Quantum*: A problem from "Where Do Problems Come From?" by I. Sharygin, Jan./Feb. 2001, pages 18–28)

In $\triangle ABC$, angle B is $120°$ (Figure 16). BM is the bisector of angle B and the bisector CK of exterior angle BCD meets AB extended at K. KM crosses BC at P. How big is $\triangle APM$?

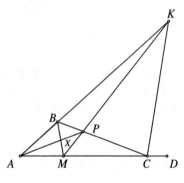

FIGURE 16

In $\triangle ABC$, each half of $\angle B$ is $60°$, and the exterior angle CBK is also $60°$ (Figure 17). Moreover, so is $\angle EBK = 60°$. In $\triangle BMC$, then, BK and CK bisect the exterior angles at B and C, making K equidistant from all three of its sides, in particular, from MB and MC. Thus KM is the bisector of $\angle BMC$. Similarly, in $\triangle ABM$, BP and MP bisect the exterior angles at B and M, making AP the bisector of $\angle A$.

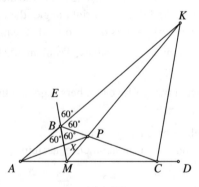

FIGURE 17

Now, in $\triangle APM$,

$$\angle APM = \text{exterior } \angle PMC - \text{interior and opposite } \angle PAM$$
$$= \tfrac{1}{2}\angle BMC - \tfrac{1}{2}\angle A$$
$$= \tfrac{1}{2}(\angle BMC - \angle A).$$

In $\triangle ABM$, $\angle BMC$ is the exterior angle at M and $\angle A$ is the interior and opposite angle. Hence

$$\angle BMC - \angle A = (\text{the other interior angle) } \angle ABM = 60°,$$

and we have $\angle APM = \tfrac{1}{2} \cdot 60° = 30°$.

24. (*Quantum*: From "Problems Teach Us How To Think" by V. Proizvolov, Jan./Feb. 2002, pages 42–45)

    $ABCDE$ is a zigzag line in a circle such that the angles at B, C, and D are each $45°$ (Figure 18(a)). Prove that

$$AB^2 + CD^2 = BC^2 + DE^2.$$

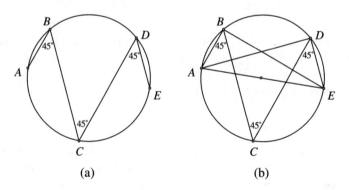

(a)   (b)

FIGURE 18

    If an arc subtends a $45°$ angle at the circumference, it subtends a right angle at the center, and its length is therefore one-quarter of the circumference. Hence arc ACE is one-half the circumference, making AE a diameter (Figure 18(b)). Thus angles ABE and ADE are right angles, and the theorem of Pythagoras gives

$$AB^2 + BE^2 = AD^2 + DE^2 \quad (= AE^2).$$

Since each of the arcs AC, BD, CE subtend the same angle at the circumference, they are equal arcs. Hence

$$\text{arc } CAB = \text{arc } ABD,$$

implying $BC = AD$. Similarly, arcs BDE and DEC are equal, and $BE = CD$. Therefore

$$AB^2 + BE^2 = AD^2 + DE^2$$

gives

$$AB^2 + CD^2 = BC^2 + DE^2.$$

25. (*Quantum*: From the article "Taking on Triangles" by A. Kanel and A. Ko-valdzhi, Mar./Apr. 2001, pages 10–16)

> Ali Baba arrived at a cave where there are gold, diamonds, and a trunk. If the trunk is filled with gold, it weighs 200 kilograms; if it is filled with diamonds, it weights 40 kilograms; the empty trunk is weightless. A kilogram of gold is worth 20 dinars and a kilogram of diamonds is worth 60 dinars. What is the greatest value of the treasure that Ali can take away if he can carry only 100 kilograms?

Let x and y denote the number of kilograms of gold and diamonds, respectively, that are packed into the trunk. Then (x, y) is a point in the closed first quadrant of a coordinate plane.

Since Ali can carry only 100 kilograms, then

$$x + y \leq 100. \tag{1}$$

If we let the volume of the trunk be 200 units, then a kilogram of gold would occupy $\frac{200}{200} = 1$ unit of volume and a kilogram of diamonds would occupy $\frac{200}{40} = 5$ units, and x and y would also be constrained by

$$x + 5y \leq 200. \tag{2}$$

These restraints confine (x, y) to the closed shaded region in Figure 19. We want to find the point (x, y) in this region that maximizes $20x + 60y$. To this end, consider the family of straight lines $x + 3y = k$. When $x + 3y$ is a maximum, so is

$$20(x + 3y) = 20x + 60y = 20k.$$

Now, $x + 3y = k$ is the family of straight lines with slope $-\frac{1}{3}$, a slope that is steeper than the slope $-\frac{1}{5}$ of $x + 5y = 200$ but not as steep

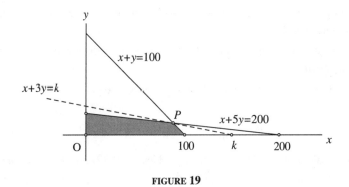

FIGURE **19**

as the slope -1 of $x + y = 100$, and the x-intercept k of a member will increase the farther the line is from the origin. Therefore, if a straight line of slope $-\frac{1}{3}$ is swept across the first quadrant from the origin, it is clear from the figure that it will cease to make contact with the shaded region when it gets beyond the point of intersection $P(75, 25)$ of the leading edges of the half-planes in (1) and (2). Hence k is a maximum for the member through P, where its value is

$$75 + 3(25) = 150,$$

and the maximum value of the treasure Ali can take away is $20(150) = 3000$ dinars.

I know what you're thinking—why bother with all this when $(x, y) = (75, 25)$ can be found right away by solving $x + y = 100$ and $x + 5y = 200$? Surely, to get the greatest value of treasure, you wouldn't leave any empty space in the trunk and you would cart off as much as you could possibly carry, implying that (1) and (2) should be equations, not inequalities.

It's not inconceivable, however, that the greatest value might be given by some unexpected combination of gold and diamonds that doesn't fill the trunk and/or doesn't weigh the whole 100 kilograms. The only way to be sure is to deal with inequalities in (1) and (2).

To illustrate this point, suppose the value of gold is only 10 dinars per kilogram. In this case the sweep line would be $x + 6y = k$ ($10x + 60y$ is a maximum when $x + 6y$ is), a line of slope $-\frac{1}{6}$, which is not as steep as the slope $-\frac{1}{5}$ of $x + 5y = 200$. Hence the last point of contact of the sweep line and the shaded region would be the y-intercept $(0, 40)$ of $x + 5y = 200$, implying that Ali should forget the gold and fill the trunk with diamonds, even though it would only weigh 40 kilograms. This would give him a maximum of 2400 dinars.

26. Prove the surprising result that $a^{1/\ln a}$ is a constant for all positive real numbers $a \neq 1$.

Let $y = a^{1/\ln a}$. Then, taking logarithms, we get

$$\ln y = \frac{1}{\ln a} \cdot \ln a = 1, \quad \text{and} \quad y = e.$$

27. (A Problem of Rex H. Wu, Brooklyn, New York (*The College Mathematics Journal*, March 2003, page 115))

Use Figure 20 to verify each of the equations

(a) $\tan^{-1}(1/2) + \tan^{-1}(1/3) = \pi/4$,

(b) $\tan^{-1}(3) - \tan^{-1}(1/2) = \pi/4$,

(c) $\tan^{-1}(2) - \tan^{-1}(1/3) = \pi/4$,

(d) $\tan^{-1}(1) + \tan^{-1}(1/2) + \tan^{-1}(1/3) = \pi/2$,

(e) $\tan^{-1}(1) + \tan^{-1}(2) + \tan^{-1}(3) = \pi$.

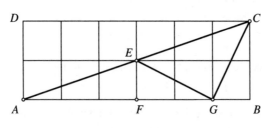

FIGURE 20

Right triangles EFG and GBC are congruent (SAS), implying

$$EG = GC, \quad \text{and} \quad \angle FEG = \angle CGB.$$

Since the angles at E and G in right triangle EFG add to a right angle, then so do angles EGF and CGB, implying $\angle EGC$ is a right angle and that triangle EGC is an isosceles right triangle. Hence in $\triangle EGC$ (Figure 21), $s + t = u = \pi/4$.

Let the angles v, x, y, z be as marked in Figure 21. Then we have

$$v = \tan^{-1}(1/2), \qquad x = \tan^{-1}(1/3),$$
$$y = \tan^{-1}(2), \qquad z = \tan^{-1}(3),$$

and since $\angle CGB = \angle DCG = x + u$,

$$\tan^{-1}(2) = x + u.$$

Also,

$$\tan^{-1}(3) = z = u + v.$$

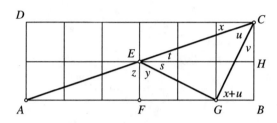

FIGURE 21

Then

(a) $\tan^{-1}(1/2) + \tan^{-1}(1/3) = s + t = \pi/4,$

(b) $\tan^{-1}(3) - \tan^{-1}(1/2) = (u + v) - v = u = \pi/4,$

(c) $\tan^{-1}(2) - \tan^{-1}(1/3) = x + u - x = u = \pi/4,$

(d) $\tan^{-1}(1) + \tan^{-1}(1/2) + \tan^{-1}(1/3) = u + v + x = \angle DCB = \pi/2,$

(e) $\tan^{-1}(1) + \tan^{-1}(2) + \tan^{-1}(3) = (s + t) + y + z = \angle AEC = \pi.$

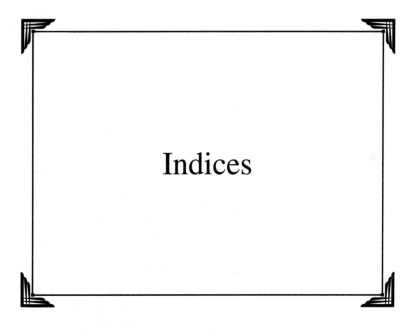

Indices

Index of Publications

The section(s) in which the publication is mentioned is given immediately after its name.

1. *Mathematical Miniatures*, Titu Andreescu and Svetoslav Savchev (MAA, Anneli Lax New Mathematical Library Series, 2003): section 1.
2. *Leningrad Mathematical Olympiads, 1987–1991*, Dimitry Fomin and Alexey Kirichenko (MathPro Press, 1994): section 9.
3. *The Contest Problem Book VI*, Leo Schneider (MAA, the Anneli Lax New Mathematical Library Series, 2000): section 2.
4. *Mathematics Magazine* (an MAA journal): section 4, Challenges 5, 6.
5. *The College Mathematics Journal* (an MAA journal): section 5, Challenges 16, 17, 27.
6. *The Contest Problem Book V*, George Berzsenyi and Stephen Maurer (MAA, The Anneli Lax New Mathematical Library Series, 1997): section 10.
7. *Problem-Solving Through Problems*, Loren Larson (Springer-Verlag, 1983): section 3.
8. *The New Mexico Mathematics Contest Problem Book*, Liong-shin Hahn, (University of New Mexico Press, forthcoming): section 8.
9. *Problems in Plane Geometry*, Viktor Prasolov (unpublished in English): section 7.
10. *Quantum*, (a magazine, now defunct, formerly published by The National Science Teachers Association in cooperation with the Quantum Bureau of the Russian Academy of Sciences): section 11, Challenges 1, 2, 3, 9, 10, 21, 22, 23, 24, 25.
11. *Pi Mu Epsilon Journal*, (The National Honorary Mathematics Society): section 6, Challenges 18, 19, 20.
12. *Problems and Solutions From The Mathematical Visitor*, edited by Stanley Rabinowitz, (MathPro Press, 1996) section 12.

13. *Complex Numbers and Geometry*, Liong-shin Hahn (MAA, Spectrum Series, 1990): section 13.

14. *In Polya's Footsteps*, Ross Honsberger, MAA, Dolciani Series, 1997): section 13.

15. *From Erdős to Kiev*, Ross Honsberger, (MAA, Dolciani Series, 1996): section 13.

16. *The Penguin Dictionary of Curious and Interesting Geometry*, David Wells, (Penguin, 1991): section 17.

17. *San Gaku*, Hidetosi Fukagawa and Dan Pedoe, (The Charles Babbage Research Centre, Winnipeg, Canada): section 17.

18. *The Book of Prime Number Records*, Paulo Ribenboim, (Springer-Verlag, second edition, 1989): section 18.

19. *Mathematical Olympiad Problems*, Titu Andreescu and Razvan Gelca, (Birkhuser Boston, 2000): Challenges 12, 13.

Subject Index

It is hoped that these brief descriptions will identify a topic sufficiently to redirect you to its section.

Algebra, Number Theory, and Probability   Section

Index of Names

The numbers following a name list the sections in which the name appears.

General Index

The number(s) after the item give its section(s) in the text.